OPPOSING
VIEWPOINTS®
SERIES

D1449056

Space Exploration
and the Search for
Life Beyond Earth

Other Books of Related Interest

Opposing Viewpoints Series

Artificial Intelligence and the Technological Singularity
The Future of Space Exploration
The US Military

At Issue Series

The Federal Budget and Government Spending
Food Security
Open Borders

Current Controversies Series

Environmental Catastrophe
The Political Elite and Special Interests
Sustainable Consumption

> "Congress shall make no law … abridging the freedom of speech, or of the press."

First Amendment to the US Constitution

The basic foundation of our democracy is the First Amendment guarantee of freedom of expression. The Opposing Viewpoints series is dedicated to the concept of this basic freedom and the idea that it is more important to practice it than to enshrine it.

OPPOSING
VIEWPOINTS®
SERIES

Space Exploration and the Search for Life Beyond Earth

Avery Elizabeth Hurt, Book Editor

GREENHAVEN
PUBLISHING

Published in 2022 by Greenhaven Publishing, LLC
353 3ʳᵈ Avenue, Suite 255, New York, NY 10010

Articles in Greenhaven Publishing anthologies are often edited for length to meet page
requirements. In addition, original titles of these works are changed to clearly present
the main thesis and to explicitly indicate the author's opinion. Every effort is made to
ensure that Greenhaven Publishing accurately reflects the original intent of the authors.
Every effort has been made to trace the owners of the copyrighted material.

Cover image: Triff/Shutterstock.com

Library of Congress Cataloging-in-Publication Data

Names: Hurt, Avery Elizabeth, editor.
Title: Space exploration and the search for life beyond Earth / Avery
Elizabeth Hurt.
Description: First edition. | New York : Greenhaven Publishing, 2022. |
Series: Opposing viewpoints | Includes bibliographical references and
index. | Contents: Alien life might already live among us / Samantha
Rolfe -- Aliens may have already visited / Jasper Hamill -- Advanced
civilizations have probably developed previously elsewhere in the
universe / Lenor Sierra -- Most likely number of contactable alien
civilisations is 36 / Nicola Davis -- The aliens we're looking for may
already be dead / Tom Siegfried. | Audience: Ages 15+ | Audience: Grades
10-12 | Summary: "Anthology of essays exploring attempts to understand
life beyond Earth"-- Provided by publisher.
Identifiers: LCCN 2021037225 | ISBN 9781534508415 (library binding) | ISBN
9781534508408 (paperback)
Subjects: LCSH: Life on other planets--Juvenile literature. |
Extraterrestrial beings--Juvenile literature. | Outer
space--Exploration--Juvenile literature.
Classification: LCC QB54 .S684 2022 | DDC 576.8/39--dc23/eng/2021101
LC record available at https://lccn.loc.gov/2021037225

Manufactured in the United States of America

Website: http://greenhavenpublishing.com

Contents

Chapter 3: Is It Foolhardy to Send Out Signals Letting Potential Enemies Know We Are Here?

Chapter 4: Is the Search for Life Elsewhere in the Universe a Waste of Resources?

The Importance of Opposing Viewpoints

Perhaps every generation experiences a period in time in which the populace seems especially polarized, starkly divided on the important issues of the day and gravitating toward the far ends of the political spectrum and away from a consensus-facilitating middle ground. The world that today's students are growing up in and that they will soon enter into as active and engaged citizens is deeply fragmented in just this way. Issues relating to terrorism, immigration, women's rights, minority rights, race relations, health care, taxation, wealth and poverty, the environment, policing, military intervention, the proper role of government—in some ways, perennial issues that are freshly and uniquely urgent and vital with each new generation—are currently roiling the world.

If we are to foster a knowledgeable, responsible, active, and engaged citizenry among today's youth, we must provide them with the intellectual, interpretive, and critical-thinking tools and experience necessary to make sense of the world around them and of the all-important debates and arguments that inform it. After all, the outcome of these debates will in large measure determine the future course, prospects, and outcomes of the world and its peoples, particularly its youth. If they are to become successful members of society and productive and informed citizens, students need to learn how to evaluate the strengths and weaknesses of someone else's arguments, how to sift fact from opinion and fallacy, and how to test the relative merits and validity of their own opinions against the known facts and the best possible available information. The landmark series Opposing Viewpoints has been providing students with just such critical-thinking skills and exposure to the debates surrounding society's most urgent contemporary issues for many years, and it continues to serve this essential role with undiminished commitment, care, and rigor.

The key to the series's success in achieving its goal of sharpening students' critical-thinking and analytic skills resides in its title—

Opposing Viewpoints. In every intriguing, compelling, and engaging volume of this series, readers are presented with the widest possible spectrum of distinct viewpoints, expert opinions, and informed argumentation and commentary, supplied by some of today's leading academics, thinkers, analysts, politicians, policy makers, economists, activists, change agents, and advocates. Every opinion and argument anthologized here is presented objectively and accorded respect. There is no editorializing in any introductory text or in the arrangement and order of the pieces. No piece is included as a "straw man," an easy ideological target for cheap point-scoring. As wide and inclusive a range of viewpoints as possible is offered, with no privileging of one particular political ideology or cultural perspective over another. It is left to each individual reader to evaluate the relative merits of each argument—as he or she sees it, and with the use of ever-growing critical-thinking skills—and grapple with his or her own assumptions, beliefs, and perspectives to determine how convincing or successful any given argument is and how the reader's own stance on the issue may be modified or altered in response to it.

This process is facilitated and supported by volume, chapter, and selection introductions that provide readers with the essential context they need to begin engaging with the spotlighted issues, with the debates surrounding them, and with their own perhaps shifting or nascent opinions on them. In addition, guided reading and discussion questions encourage readers to determine the authors' point of view and purpose, interrogate and analyze the various arguments and their rhetoric and structure, evaluate the arguments' strengths and weaknesses, test their claims against available facts and evidence, judge the validity of the reasoning, and bring into clearer, sharper focus the reader's own beliefs and conclusions and how they may differ from or align with those in the collection or those of their classmates.

Research has shown that reading comprehension skills improve dramatically when students are provided with compelling, intriguing, and relevant "discussable" texts. The subject matter of these collections could not be more compelling, intriguing, or urgently relevant to today's students and the world they are poised to inherit. The anthologized articles and the reading and discussion questions that are included with them also provide the basis for stimulating, lively, and passionate classroom debates. Students who are compelled to anticipate objections to their own argument and identify the flaws in those of an opponent read more carefully, think more critically, and steep themselves in relevant context, facts, and information more thoroughly. In short, using discussable text of the kind provided by every single volume in the Opposing Viewpoints series encourages close reading, facilitates reading comprehension, fosters research, strengthens critical thinking, and greatly enlivens and energizes classroom discussion and participation. The entire learning process is deepened, extended, and strengthened.

For all of these reasons, Opposing Viewpoints continues to be exactly the right resource at exactly the right time—when we most need to provide readers with the critical-thinking tools and skills that will not only serve them well in school but also in their careers and their daily lives as decision-making family members, community members, and citizens. This series encourages respectful engagement with and analysis of opposing viewpoints and fosters a resulting increase in the strength and rigor of one's own opinions and stances. As such, it helps make readers "future ready," and that readiness will pay rich dividends for the readers themselves, for the citizenry, for our society, and for the world at large.

Introduction

> *"Nothing in the universe is unique and*
> *alone and therefore in other regions*
> *there must be other earths inhabited*
> *by different tribes of men and breeds*
> *of beasts."*
>
> *—Lucretius, Roman*
> *Epicurean poet*

A re we alone in the universe? Humans have probably always looked up at the heavens and wondered if there was anyone else out there. The ancient Greeks certainly wondered. Democritus, a Greek scholar who lived in the fifth century BCE, was the first to suggest that everything that exists is made of the same tiny particles, just arranged in a myriad of ways. Epicurus came along a couple of hundred years later and took up Democritus's idea—and the implications of it. Epicurus wrote:

> There is an infinite number of worlds, some like this world, others unlike it...For the atoms out of which a world might arise, or by which a world might be formed, have not all been expended on one world or a finite number of worlds, whether like or unlike this one. Hence there will be nothing to hinder an infinity of worlds.

Lucretius, a Roman poet who came along around 99–55 BCE and helped to spread Epicurus's ideas, put it more bluntly: "Earth is not the only planet with life."

But only in the last half-century have we had the technology to look for that life. In 1959, ten years before Neil Armstrong took "one small step" onto the surface of the moon, Frank Drake (an astronomer you will hear more about in this volume) conducted the first organized search for life elsewhere in the universe. Then in 1984, Drake and others established the SETI Institute. "SETI" stands for Search for Extra Terrestrial Intelligence, and search is exactly what the scientists at SETI do. At first, this was a passive endeavor—scanning the skies with radio telescopes trying to pick up signals from any civilizations that might be advanced enough to send them. Since then, we've advanced to sending messages to any life forms that might be there to hear them.

Here? There? Where?

While scientists search for life on faraway planets, some people think they're looking in the wrong places. Perhaps aliens are already here, orbiting Earth in spaceships. Believing in UFOs is still a fringe idea, right up there with ghosts and ESP. But after all, "UFO" simply stands for "unidentified flying object," and it turns out there are far more unidentified objects zipping around in Earth's atmosphere than most people realize. The US government recently admitted as much but (presumably to avoid being labeled as nutters) renamed them "unidentified aerial phenomena" (UAP). For some years now, military pilots have been seeing what look like aircraft doing things that no known aircraft is capable of doing. Theories about the strange, zipping lights seen in Earth's atmosphere range from faulty equipment to new Chinese spy technology. A government report on UAP was released in 2021, and you can read it in full in chapter 2. The report's conclusions were that no one knows what's behind the odd sightings made by military pilots. Aliens are unlikely but can't be totally ruled out. This fits well with the tone of the viewpoints contained in this volume. The question of whether there is life other than us in the universe is serious and the answer is complex. In this resource, you will hear from a variety of people, many of them scientists, debating this and related questions. They disagree on quite a lot. The one thing they do agree on, however, is that the question is a valid one.

Serious Questions, Complex Answers

Opposing Viewpoints: Space Exploration and the Search for Life Beyond Earth attempts to understand our place in the universe and whether or not humankind should expand its knowledge of it. In chapters titled "What Are the Chances That There Is Intelligent Life Elsewhere in the Universe?" "Has Earth Already Been Visited by Extraterrestrial Life?" "Is It Foolhardy to Send Out Signals Letting Potential Enemies Know We Are Here?" and "Is the Search for Life Elsewhere in the Universe a Waste of Resources?," viewpoint authors tackle our fascination with solving the mysteries of the universe from a variety of angles.

In the first chapter, the authors address the question, What are the odds? You'll learn about the Drake equation (named for Frank Drake), which tries to get a mathematical grasp on the likelihood that we aren't alone, and the Fermi paradox, which points out the peculiar fact that despite the relatively high probability of life elsewhere (according to many estimates), we still haven't heard from anyone. Does that mean that there is, in fact, no one out there? Chapter 2 features authors who consider the possibility that Earth has *already* been visited by aliens. In Chapter 3, the authors ask, Is it dangerous to reach out to potential alien life? Some think we should be wary, others say the benefits are worth the risk, and some declare it a moot question. If advanced alien life is out there, they already know about us. In the last chapter, the viewpoints turn to a more pragmatic issue: Is space exploration and the search for alien life worth the money? Should we spend that money on the many problems still to solve here on Earth? Or does the science we learn from the search give us the tools we need to improve human life, maybe even save it? None of these questions have easy answers, but the discussion of all of them is both fascinating and thought provoking.

What Are the Chances That There Is Intelligent Life Elsewhere in the Universe?

Chapter Preface

It's fun to speculate about the chances that there is life elsewhere in the universe. People have been doing it for most of human history. And for most of human history, speculation was all we could do. But in recent years, scientists have brought their entire toolkit, including mathematics, physics, cosmology, and evolutionary biology, to bear on the question. We may not have found extraterrestrial life yet, but we're definitely making progress in sorting out where that life might or might not be and what are the odds that it exists at all.

The authors represented in this chapter—both science journalists and working scientists—explore the question "Are we alone?" from a variety of viewpoints, some of them quite fresh and surprising. In one article, the author and the scientists she speaks with look at how new scientific discoveries and methods make it possible to make more informed estimates of the chances of alien life than was possible even a few decades ago. The question is still very much an open one, but at least the discussion is beginning to be more data-based and less speculative.

In another viewpoint, people at work on the problem roll up their sleeves, crunch the data, and estimate that there are likely to be 36—an oddly specific number—technologically advanced civilizations in our galaxy.

Meanwhile, other writers wonder if a civilization sufficiently advanced to be able to develop interplanetary communication (much less travel) has any hope of lasting long enough to make that a reality. It's unlikely, they argue, that a civilization would last as long as it takes to visit a faraway planet.

Another viewpoint looks closer to home and starts smaller. The life we're looking for may be just under our noses, though it may not be anything like us at all—and that could make it almost impossible to recognize, even if we do see it.

> *"Our Milky Way galaxy has plenty of stars, plenty of planets, and plenty of time to develop intelligent lifeforms. Yet so far, we've seen no sign of such technology, nor heard a peep of conversation."*

The Ingredients for Earthly Life Appear to Be Almost Everywhere We've Looked

Pat Brennan

In the following viewpoint, Pat Brennan writing for NASA, the US National Aeronautics and Space Administration, explains that whether or not there is intelligent life elsewhere in the universe isn't known. However, the odds are improving that we will find it if it's out there, thanks to advances in technology and the likelihood that we can develop and launch a space telescope powerful enough to detect it. Pat Brennan is a science writer for NASA's Exoplanet Exploration Program and the NASA Sea Level Portal.

As you read, consider the following questions:

1. What is the habitable zone?
2. Why is water so important in the search for extraterrestrial life?
3. What sorts of signatures of alien life and alien intelligence are astronomers looking for?

"Life in the Universe: What Are the Odds?" by Pat Brennan, NASA, March 9, 2021.

As humanity casts an ever-wider net across the cosmos, capturing evidence of thousands of worlds, an ancient question haunts us: Is anybody out there?

The good news: We know vastly more than any previous generation. Our galaxy is crowded with exoplanets—planets around other stars. A healthy percentage of them are small, rocky worlds, of a similar size and likely similar composition to our home planet.

The ingredients in the recipe for earthly life—water, elements associated with life, available sources of energy—appear to be almost everywhere we've looked.

Now the bad news. We have yet to find another "Earth" with life, intelligent or not. Observing signs of possible microbial life in exoplanet atmospheres is currently just out of reach. No convincing evidence of advanced technology—artificial signals by radio or other means, or the telltale sign of, say, massive extraterrestrial engineering projects—has yet crossed our formidable arrays of telescopes in space or on the ground.

And finding non-intelligent life is far more likely; Earth existed for most of its history, 4.25 billion years, without a whisper of technological life, and human civilization is a very late-breaking development.

Is there life beyond Earth? So far, the silence is deafening.

"I hope it's there," said Shawn Domagal-Goldman, a research astronomer at NASA's Goddard Space Flight Center in Greenbelt, Maryland. "I want it to be there. I'll be planning a party if we find it."

Domagal-Goldman co-leads a team of exoplanet hunters who, in the years and decades ahead, are planning to do just that. Working with scientists across NASA, as well as academic and international partners, his team and others are helping to design and build the next generation of instruments to sift through light from other worlds, and other suns. The goal: unambiguous evidence of another living, breathing world.

While the chances of finding life elsewhere remain unknown, the odds can be said to be improving. A well-known list of the data needed to determine the likely abundance of life-bearing worlds, though highly conjectural, is known as the "Drake equation."

Put forward in 1961 by astronomer Frank Drake, the list remains mostly blank. It begins with the rate of star formation in the galaxy and the fraction of stars that have planets, leading step-by-step through the portion of planets that support life and—most speculatively—to the existence and durability of detectable, technological civilizations.

When Drake introduced this roadmap to life beyond Earth, all the terms—the signposts along the way—were blank.

Some of the first few items are now known, including the potential presence of habitable worlds, said researcher Ravi Kopparapu from Goddard, also a co-leader of Domagal-Goldman's team. He studies the habitability and potential for life on exoplanets.

If we develop and launch a powerful enough space telescope, "we could figure out if we have advanced life or biological life," he said.

Finding a Planet That's "Just Right"

Drake's list can be a good conversation starter, and a useful way to frame the complex questions around the possibility of other life. But these days, scientists don't spend a great deal of time discussing it, Domagal-Goldman said.

Instead, they use a narrower yardstick: the habitable zone.

Every star, like every campfire, has a definable zone of radiated warmth. Too close, and your marshmallow—or your planet—might end up as nothing more than a charred cinder. Too far away, and its surface remains cold and unappetizing.

In both cases, "just right" is more likely to be somewhere in between.

For a planet, the habitable zone is the distance from a star that allows liquid water to persist on its surface—as long as that planet has a suitable atmosphere.

In our solar system, Earth sits comfortably inside the Sun's habitable zone. Broiling planet Venus is within the inner edge, while refrigerated Mars is near the outer boundary.

Determine the distance of an exoplanet from the star itself, as well as the star's size and energy output, and you can estimate whether the planet falls within the habitable zone.

For larger, hotter stars, the zone is farther away; for smaller, cooler stars, it can be very close indeed. Finding these "just right" planets in the habitable zone is one of the keys to finding signs of life.

"If they fit within these parameters, they could potentially support a temperate environment," said Natasha Batalha, a research scientist at the NASA Ames Research Center. "Therefore it would be incredibly interesting to study their atmospheres."

Batalha's specialty, in fact, is finding ways to read exoplanet atmospheres—and building computer models to better understand them.

"That is the next step, the next frontier," she said.

The habitable zone concept is not yet definitive. Small, rocky worlds like ours that orbit other stars are too far away to determine whether they have atmospheres, at least using present-day technology.

That's where teams like the one co-led by Kopparapu and Domagal-Goldman come in. The space telescopes and instruments now on their drawing boards are meant to be powerful enough to peer into these atmospheres and identify the molecules present. That will tell us which gases dominate.

We could find a small, rocky, watery world around a Sun-like star with an atmosphere of nitrogen, oxygen, and carbon dioxide: a little like looking in a mirror.

"To search for life anywhere, we have this 'follow the water' approach," Domagal-Goldman said. "Anywhere you find water on Earth, you find life. Whether it's life on Mars, ocean worlds, or exoplanets, water is the first signpost we're looking for."

For now, the habitable zone remains a kind of first cut in the search for life-bearing worlds.

"The habitable zone is a very useful tool for mission design," said Rhonda Morgan of NASA's Jet Propulsion Laboratory in Southern California.

She studies how to use the data gathered so far on exoplanets to refine designs for future space telescopes.

Over the past quarter century, thousands of exoplanets have been confirmed in a Milky Way galaxy that likely holds trillions. Thousands more will come to light in the years ahead. Tools like the habitable zone will help planet hunters sort through these growing ranks to pick the most likely candidates for supporting life.

"We are in a position now where we can propose a potential, future mission that would be capable of directly imaging an Earth-like planet around a nearby, Sun-like star," she said. "This is the first time in history that the technology has been this close, probably less than 10 years from launch."

Still, we might need something beyond the habitable-zone concept for more extreme cases. It won't help much, for instance, with "weird" life—life as we don't know it. Living things on other worlds might use vastly different chemistry and molecular compounds, or even a solvent other than water.

"This is one of the questions we get from the public often: If there are aliens, how are we going to recognize them if they're really weird?" Domagal-Goldman said. "How do we find what we would consider to be weird life? And how do we make sure not to be tricked by strange, dead planets that look alive—mirages in the desert?"

Life on planets around other stars also might be hidden in a subsurface ocean encased in ice, invisible even to our most powerful space telescopes. Moons of Jupiter and Saturn are known to harbor such oceans, some revealing through remote sensing at least a few of the characteristics we expect for habitable worlds.

Some "exo-moons" also might be habitable worlds, as in the film *Avatar*. But even proposed, future instruments are unlikely to have sufficient power to detect atmospheres of moons around giant exoplanets.

Still, the habitable zone is a good start, a way to zero in on signs of life made familiar by our fellow organisms here on planet Earth.

Cosmic Eavesdropping

A shortcut to finding lifeforms like ourselves, of course, would be to intercept tech-savvy communications. Searches for signs of intelligent life have been underway for decades.

In recent years, among NASA scientists, such potential signs have acquired an intriguing new name: technosignatures.

Evidence of a communicative, technological species somewhere among the endless fields of stars could come in the traditional form: signals by radio or optical light waves, or from some other slice of the electromagnetic spectrum.

But scientists imagine many other forms. An exoplanet atmosphere might show signs of synthetic gases, such as CFCs, revealing an industrial species like us.

Or maybe we'll see the glimmer of something like a "Dyson sphere," popularized by physicist Freeman Dyson: an epic-scale structure built around a star to capture the lion's share of its energy.

Such possibilities remain speculative. For now, we have no real answer to a disturbing question from another 20th century physicist, Enrico Fermi.

Where is everybody?

The question has fueled more than 70 years of debate, but boils down to a simple observation. Our Milky Way galaxy has plenty of stars, plenty of planets, and plenty of time to develop intelligent lifeforms—some of whom might well have had billions of years to develop interstellar travel.

Yet so far, we've seen no sign of such technology, nor heard a peep of conversation. Why is the cosmos so profoundly silent?

"If life had so much time to evolve, why haven't we found it?" Batalha asks, to summarize the question. "Why isn't life just crawling everywhere in the galaxy, or the universe? It could be a combination of a lot of things. Space travel is very difficult for us."

Vast amounts of energy would be needed just to get us to our nearest neighboring star, Proxima Centauri, she said. "It would just be incredibly expensive, and require a lot of resources."

And once we—or some other civilization—reached such a distant destination, she said, another problem would be perpetuating the travelers' existence into future generations.

Experts offer many reasons why somebody, or something, might be out there, yet beyond our detection. On the other hand, the ultra-cautious might remind us that, while a lifeless cosmos seems unlikely, we have exactly zero information one way or the other.

Still, scientists like Kopparapu say they like our chances of finding some form of life, and are hard at work on the telescopes and instruments that could make that future, party-starting epiphany a reality.

"It's not a question of 'if,' it's a question of 'when' we find life on other planets," he said. "I'm sure in my lifetime, in our lifetime, we will know if there is life on other worlds."

> "We have limited ways of studying
> the microscopic world as only a
> small percentage of microbes can be
> cultured in a lab. This may mean
> that there could indeed be many
> lifeforms we haven't yet spotted."

We Might Not Recognize Alien Life
Even If It's Right Under Our Feet

Samantha Rolfe

*Most discussions about the possibility of alien life assume two
things: first, that we'll recognize that life if we see it, and second,
that if it exists at all, it exists way off on another faraway planet.
In the following viewpoint, Samantha Rolfe questions both of those
assumptions. Alien life, she argues, might actually exist in a "shadowy
biosphere" that we can't see because we can't even comprehend it.
In fact, the intelligent life we're searching for doesn't even have
to be humanoid. Our assumptions about what we're looking for
might in actuality be limiting our ability to see it. Samantha Rolfe
is a lecturer in astrobiology and principal technical officer at the
Bayfordbury Observatory at the University of Hertfordshire in the
United Kingdom.*

"Could Invisible Aliens Really Exist Among Us? An Astrobiologist Explains," by Samantha
Rolfe, The Conversation, January 10, 2020, https://theconversation.com/could-invisible
-aliens-really-exist-among-us-an-astrobiologist-explains-129419. Licensed under CC BY-4.0.

As you read, consider the following questions:

1. Why, according to the viewpoint, is the difficulty in defining life especially problematic when it comes to the search for alien life?
2. What is the difference between "alien" and "unfamiliar," and how does that distinction apply to the argument being made here?
3. If Earth harbors life from other planets, how did it get here, according to the author?

Life is pretty easy to recognise. It moves, it grows, it eats, it excretes, it reproduces. Simple. In biology, researchers often use the acronym "MRSGREN" to describe it. It stands for movement, respiration, sensitivity, growth, reproduction, excretion and nutrition.

But Helen Sharman, Britain's first astronaut and a chemist at Imperial College London, recently said that alien lifeforms that are impossible to spot may be living among us. How could that be possible?

While life may be easy to recognise, it's actually notoriously difficult to define and has had scientists and philosophers in debate for centuries—if not millennia. For example, a 3D printer can reproduce itself, but we wouldn't call it alive. On the other hand, a mule is famously sterile, but we would never say it doesn't live.

As nobody can agree, there are more than 100 definitions of what life is. An alternative (but imperfect) approach is describing life as "a self-sustaining chemical system capable of Darwinian evolution," which works for many cases we want to describe.

The lack of definition is a huge problem when it comes to searching for life in space. Not being able to define life other than "we'll know it when we see it" means we are truly limiting ourselves to geocentric, possibly even anthropocentric, ideas of

DNA SEQUENCING COULD IDENTIFY ALIEN LIFE

Here's a riddle: If an alien life form is, well, alien, how will we know what it is? DNA and RNA are the building blocks of life on Earth, but the molecules of life might differ substantially on another planet. So if scientists combing, say, the potentially habitable waters of Jupiter's moon Europa were to stumble across a new life form, how could they know what they had discovered?

A new paper by scientists at Georgetown University, published online this month in the journal *Astrobiology*, suggests a method for identifying alien life using modern genome sequencing technology.

"Most strategies for life detection rely upon finding features known to be associated with terran life, such as particular classes of molecules," the researchers wrote. "But life may be vastly different on other planets and moons, particularly as we expand our efforts to explore ocean worlds like Europa and Enceladus."

It works like this: Nucleic acids like DNA form structures that will inherently bind to a host of materials and shapes, including organic molecules, minerals, and even metals. In the system that the researchers propose, a technique sometimes used in cancer detection called the "systematic evolution of ligands by exponential enrichment," researchers propose creating nucleic acids that can bind to organic molecules that are indicators of life. The nucleic acids would theoretically act as a sort of sensor than can be amplified, and the binding patterns analyzed, to reveal a kind of biochemical signature—a "fingerprint," as the researchers put it. The biochemistry of the alien life might be completely different from anything that we have seen on Earth, but you could still get a sense of the life form's molecular patterns and complexity, and thus a broad sense of what it is. For starters, if the molecular structures identified are complex, it's a pretty good sign that it's actually life.

"Without presupposing any particular molecular framework, this agnostic approach to life detection could be used from Mars to the far reaches of the Solar System, all within the framework of an instrument drawing little heat and power," the researchers wrote.

"How Scientists Could Use DNA Sequencing to Identify Alien Life," by Kristen V. Brown, Gizmodo, March 27, 2018.

what life looks like. When we think about aliens, we often picture a humanoid creature. But the intelligent life we are searching for doesn't have to be humanoid.

Life, but Not as We Know It

Sharman says she believes aliens exist and "there's no two ways about it." Furthermore, she wonders: "Will they be like you and me, made up of carbon and nitrogen? Maybe not. It's possible they're here right now and we simply can't see them."

Such life would exist in a "shadow biosphere." By that, I don't mean a ghost realm, but undiscovered creatures probably with a different biochemistry. This means we can't study or even notice them because they are outside of our comprehension. Assuming it exists, such a shadow biosphere would probably be microscopic.

So why haven't we found it? We have limited ways of studying the microscopic world as only a small percentage of microbes can be cultured in a lab. This may mean that there could indeed be many lifeforms we haven't yet spotted. We do now have the ability to sequence the DNA of unculturable strains of microbes, but this can only detect life as we know it—that contain DNA.

If we find such a biosphere, however, it is unclear whether we should call it alien. That depends on whether we mean "of extraterrestrial origin" or simply "unfamiliar."

Silicon-Based Life

A popular suggestion for an alternative biochemistry is one based on silicon rather than carbon. It makes sense, even from a geocentric point of view. Around 90% of the Earth is made up of silicon, iron, magnesium and oxygen, which means there's lots to go around for building potential life.

Silicon is similar to carbon, it has four electrons available for creating bonds with other atoms. But silicon is heavier, with 14 protons (protons make up the atomic nucleus with neutrons) compared to the six in the carbon nucleus. While carbon can create strong double and triple bonds to form long chains useful for many functions, such as building cell walls, it is much harder for silicon. It struggles to create strong bonds, so long-chain molecules are much less stable.

What's more, common silicon compounds, such as silicon dioxide (or silica), are generally solid at terrestrial temperatures and insoluble in water. Compare this to highly soluble carbon dioxide, for example, and we see that carbon is more flexible and provides many more molecular possibilities.

Life on Earth is fundamentally different from the bulk composition of the Earth. Another argument against a silicon-based shadow biosphere is that too much silicon is locked up in rocks. In fact, the chemical composition of life on Earth has an approximate correlation with the chemical composition of the sun, with 98% of atoms in biology consisting of hydrogen, oxygen and carbon. So if there were viable silicon lifeforms here, they may have evolved elsewhere.

That said, there are arguments in favour of silicon-based life on Earth. Nature is adaptable. A few years ago, scientists at Caltech managed to breed a bacterial protein that created bonds with silicon—essentially bringing silicon to life. So even though silicon is inflexible compared with carbon, it could perhaps find ways to assemble into living organisms, potentially including carbon.

And when it comes to other places in space, such as Saturn's moon Titan or planets orbiting other stars, we certainly can't rule out the possibility of silicon-based life.

To find it, we have to somehow think outside of the terrestrial biology box and figure out ways of recognising lifeforms that are fundamentally different from the carbon-based form. There are plenty of experiments testing out these alternative biochemistries, such as the one from Caltech.

Regardless of the belief held by many that life exists elsewhere in the universe, we have no evidence for that. So it is important to consider all life as precious, no matter its size, quantity or location. The Earth supports the only known life in the universe. So no matter what form life elsewhere in the solar system or universe may take, we have to make sure we protect it from harmful contamination—whether it is terrestrial life or alien lifeforms.

So could aliens be among us? I don't believe that we have been visited by a life form with the technology to travel across the vast distances of space. But we do have evidence for life-forming, carbon-based molecules having arrived on Earth on meteorites, so the evidence certainly doesn't rule out the same possibility for more unfamiliar life forms.

> "If civilizations have a finite duration,
> then it is possible that Earth was
> settled some time in the distant past
> and all traces of that settlement have
> been erased by geological evolution."

Aliens May Have Already Visited

Jasper Hamill

In the following viewpoint, Jasper Hamill speaks with researchers who consider the possibility that more typically imagined aliens may have visited—and even colonized—Earth long before humans evolved, leaving no trace on the planet. And more civilizations capable of visiting, or at least making contact, might still be out there, just waiting to find us. For better or worse, we're making that easier with all the light and radio signals we transmit, warn these experts. Jasper Hamill is a journalist specializing in science and technology.

As you read, consider the following questions:

1. What is the Fermi paradox, and what does it suggest about the possibility of alien life?
2. Why do the experts Hamill interviews caution against trying to understand aliens in human terms?
3. What role, according to this viewpoint, might light pollution play in making contact with alien civilizations?

"Aliens May Have Already Visited the Solar System and Settled on Earth, Scientists Say," by Jasper Hamill, Metro, September 10, 2019. Reprinted by permission.

The evidence which proves aliens are real could be lurking undiscovered beneath our feet.

That's the suggestion in a new piece of research which considers the possibility that extraterrestrial civilisations may have settled on Earth.

The study does not explore the "intent and motivation" of alien civilisations, which means its authors do not discuss whether extraterrestrials are likely to be monstrous hunter-killers hellbent on galactic destruction or peace-loving "take me to your dealer" hippies dedicated to spreading love throughout the cosmos.

It instead calculates whether extraterrestrial "exo-civilisations" could travel throughout a galaxy and goes on to consider whether they may have ever reached Earth.

However, if they did manage to find Earth and live here it could be very difficult to find traces of their settlements because all evidence would be obliterated over time.

The paper is inspired by the "Fermi Paradox," which describes the contradiction between the high likelihood of aliens existing somewhere out there in the universe and the "Great Silence" caused by the fact we've seen absolutely no sign of civilisations living anywhere but Earth.

"If civilizations have a finite duration, then it is possible that Earth was settled some time in the distant past and all traces of that settlement have been erased by geological evolution," the team wrote in an early version of their paper, which has now been peer reviewed and published in *The Astronomical Journal*.

They added: "How long ago could Earth have been (temporarily) visited or settled by such a civilization without leaving any obvious trace?

"If the settlement occurred 4 billion years ago and lasted for just 10,000 years would any record of it survive in the geological record?

"The answer is: almost certainly not. This implies a temporal horizon over which a settlement might not be 'seen.'"

There's also a chance that aliens got to our solar system and decided against staying here, the paper continued.

The authors counselled against trying to understand extraterrestrials in human terms and added: "The assumption that the Earth's life-sustaining resources make it a particularly good target for extraterrestrial settlement projects could be a naive projection onto exo-civilizations of a particular set of human attitudes that conflate expansion and exploration with conquest of (or at least indifference towards) native populations.

"One might just as plausibly posit that any extremely long-lived civilization would appreciate the importance of leaving native life and its near-space environment undisturbed."

Professor Stephen Hawking famously feared that an encounter between humanity and an alien species would be disastrous.

So you may be slightly alarmed to hear that one scientist recently suggested that extraterrestrial invaders "may already be on their way."

Jacco van Loon, an astrophysicist at Keele University, issued a rather scary warning about light pollution here on Earth.

He said that we may have already given away our location to a non-human civilisation because we light up the planet every single night with electric illuminations.

For a sense of what our first encounter with aliens might be like, we'd urge you to remember that Hawking also said extraterrestrials could be "rapacious marauders roaming the cosmos in search of resources to plunder, and planets to conquer and colonize."

"Meeting an advanced civilization could be like Native Americans encountering Columbus," he continued. "That didn't turn out so well."

In an article for the Conversation, Mr. van Loon, astrophysicist and director of Keele Observatory at Keele University, wrote: "Images of the Earth at night reveal our presence in spectacular fashion. Cities and roads outline the contours of continents, while oil platforms dot the seas and ships draw lines across the ocean. This type of light, which has replaced older, incandescent sources, is unnatural. From the orange sodium or bluish mercury lamps to white-light-emitting diodes (LEDs), the artificial origin of this 'spectrum' should be easy for technologically advanced aliens to spot.

"In the coming decades, Earth's space agencies may be developing the means to detect such artificial light from planets around other stars. But we may fail, if aliens believe the smartest thing to do is to keep quiet and remain in the dark. We, on the other hand, may already have been seen, and they may already be on their way. This begs the question—should we dim our lights, before it's too late?"

He went on to warn about the potentially disastrous effects of light pollution and the well-meaning, but potentially ruinous, decision to send radio transmissions out into space in the hope aliens will pick them up.

"Since the first use of electric lamps in the 19th century, society hasn't looked back," he continued.

Homes and streets are lit at all hours so that people can go about their business when they'd once have been asleep. Besides the obvious benefits to societies and the economy, there's growing awareness of the negative impact of artificial light.

"Light pollution has been blamed for wasting energy, disrupting wildlife behaviour and harming mental health. One aspect has avoided the spotlight though. Namely, that light not only allows one to see, but also to be seen. This could well attract unwelcome attention—and not just from moths.

"The innate curiosity of humans and our growing knowledge of the universe in which we live have led us inexorably to a question. Do civilisations exist on planets other than Earth? Scientists now believe that there are many places in the universe which might harbour simple lifeforms such as bacteria.

"What is more speculative is the notion that such extraterrestrial life could have become technologically advanced, perhaps even well beyond our capacity. This idea has captured the imagination of the general public, giving birth to a rich collection of science fiction literature and blockbuster films. But it has also received

WE'RE ALL ALONE: OXFORD STUDY SAYS CHANCE OF INTELLIGENT LIFE ELSEWHERE VERY LOW

Scientists from Oxford's Future of Humanity Institute theorize that as life evolved on earth, in many cases it depended on a series of unlikely "revolutionary transitions." Given how late intelligent life evolved on this planet, the chances of similar developments happening on other planets, before they are no longer able to sustain life, were highly unlikely, they said.

"It took approximately 4.5 billion years for a series of evolutionary transitions resulting in intelligent life to unfold on Earth," they wrote in the paper published last month. "In another billion years, the increasing luminosity of the Sun will make Earth uninhabitable for complex life.

"Together with the dispersed timing of key evolutionary transitions and plausible priors, one can conclude that the expected transition times likely exceed the lifetime of Earth, perhaps by many orders of magnitude," they wrote. "In turn, this suggests that intelligent life is likely to be exceptionally rare."

serious attention from scientists, who have thought of ways to find and possibly contact these alien species.

"In 1974, radio astronomer Frank Drake used the then most powerful radio transmitter, at Arecibo in Puerto Rico, to broadcast a message into space announcing our presence. The message will now be 45 light years away from us. While there are many stars and planets closer to us than that, they won't have been in the path of Drake's broadcast.

"But impatient as scientists tend to be, more effort has gone into searching space for such signals transmitted by extraterrestrial civilisations. As more and more planets are discovered around other stars, the search for extraterrestrial intelligence—often abbreviated to SETI—is becoming more relevant, better informed and better

To reach their conclusions, the scientists looked at statistical models to determine the probability that the sequence of evolutionary transitions that occurred on Earth could occur elsewhere.

"We made use of the assumption that what happened on Earth is typical for what happens on other planets—not the exact times, but that there are some tricky steps life needs to get through in sequence to produce intelligent observers," Oxford's Anders Sandberg told the *Daily Mail*.

The paper points to the fact that it took more than a billion years for life to advance from prokaryotic (single-cell organisms) to eukaryotes (organisms with a nucleus) means that such a step is highly unlikely.

It also notes that humans have only existed on Earth for about the last 6 million years, with *Homo sapiens* only arriving some 200,000 years ago.

"Some transitions seem to have occurred only once in Earth's history, suggesting a hypothesis reminiscent of Gould's remark that if the 'tape of life' were to be rerun, 'the chance becomes vanishingly small that anything like human intelligence' would occur," the paper says, referring to the quote from evolutionary biologist Stephen Jay Gould.

"We're All Alone: Oxford Study Says Chance of Intelligent Life Elsewhere Very Low," by Toi Staff, *Times of Israel*, December 4, 2020.

resourced. In 2015, wealthy entrepreneurs Yuri and Julia Milner allocated US$100m to the Breakthrough Listen SETI project, which buys time at observatories to use their powerful telescopes to detect artificial signals from outer space.

"Despite the vastness and emptiness of space, scientists have started to wonder why we haven't heard from aliens yet. This puzzle is known as the Fermi Paradox, named after the physicist Enrico Fermi. Among the many solutions proposed for this problem, one really brings us down to Earth: aliens might be scared of other aliens.

"While tempting, many scientists now agree that sending messages into space without knowing who might be intercepting them might not be such a good idea. Once sent, it cannot be undone.

Unlike a post on social media, it cannot be removed. Listening is much safer. But radio communication among ourselves—which includes navigation, television broadcasts and the internet—might also be detected from space.

"After all, radio waves that aren't captured continue to travel, up and away from the Earth into deep space. Unintentionally, we may already have been observed by an amused, terrified or 'interested' species, who may decide to meet us to 'shake hands', or come to enslave us, eat us, or destroy us as a precaution. We are, after all, an aggressive species ourselves.

"Fortunately, Earth has become a lot quieter, thanks to more directed signalling and fibre cables replacing aerial transmission. We might just get away with our past recklessness. But a new beacon is brightening."

> *"If we evolved in this planet, it is possible that intelligent life evolved in another part of the universe."*

The Most Likely Number of Contactable Alien Civilizations Is 36

Nicola Davis

In the following viewpoint, Nicola Davis considers the possibility that life may exist elsewhere in the universe now—not in the distant past—and that life could be in the form of advanced civilizations, not microbial life. Scientists have come up with an estimate that there could be at least 30—and possibly as many as 100—technologically advanced civilizations in our galaxy alone. However, she writes, we need to know more about our own origins before we can accurately estimate the chances that similar civilizations have evolved elsewhere. Nicola Davis is science correspondent for The Guardian.

As you read, consider the following questions:

1. Why is the Drake equation a tool for thinking about the questions rather than a formula to be solved and applied?
2. What factors does this estimate take into account that previous ones have not?
3. In what way does a civilization's distance from Earth figure into the calculations offered by these scientists?

They may not be little green men. They may not arrive in a vast spaceship. But according to new calculations there could be more than 30 intelligent civilisations in our galaxy today capable of communicating with others.

Experts say the work not only offers insights into the chances of life beyond Earth but could shed light on our own future and place in the cosmos.

"I think it is extremely important and exciting because for the first time we really have an estimate for this number of active intelligent, communicating civilisations that we potentially could contact and find out there is other life in the universe—something that has been a question for thousands of years and is still not answered," said Christopher Conselice, a professor of astrophysics at the University of Nottingham and a co-author of the research.

In 1961 the astronomer Frank Drake proposed what became known as the Drake equation, setting out seven factors that would need to be known to come up with an estimate for the number of intelligent civilisations out there. These factors ranged from the the average number of stars that form each year in the galaxy through to the timespan over which a civilisation would be expected to be sending out detectable signals.

But few of the factors are measurable. "Drake equation estimates have ranged from zero to a few billion [civilisations]—it is more like a tool for thinking about questions rather than something that has actually been solved," said Conselice.

Now Conselice and colleagues report in the *Astrophysical Journal* how they refined the equation with new data and assumptions to come up with their estimates.

"Basically, we made the assumption that intelligent life would form on other [Earth-like] planets like it has on Earth, so within a few billion years life would automatically form as a natural part of evolution," said Conselice.

The assumption, known as the Astrobiological Copernican Principle, is fair as everything from chemical reactions to star formation is known to occur if the conditions are right, he said. "[If intelligent life forms] in a scientific way, not just a random way or just a very unique way, then you would expect at least this many civilisations within our galaxy," he said.

He added that, while it is a speculative theory, he believes alien life would have similarities in appearance to life on Earth. "We wouldn't be super shocked by seeing them," he said.

Under the strictest set of assumptions—where, as on Earth, life forms between 4.5bn and 5.5bn years after star formation—there are likely between four and 211 civilisations in the Milky Way today capable of communicating with others, with 36 the most likely figure. But Conselice noted that this figure is conservative, not least as it is based on how long our own civilisation has been sending out signals into space—a period of just 100 years so far.

The team add that our civilisation would need to survive at least another 6,120 years for two-way communication. "They would be quite far away … 17,000 light years is our calculation for the closest one," said Conselice. "If we do find things closer … then that would be a good indication that the lifespan of [communicating] civilisations is much longer than a hundred or a few hundred years, that an intelligent civilisation can last for thousands or millions of years. The more we find nearby, the better it looks for the long-term survival of our own civilisation."

Dr. Oliver Shorttle, an expert in extrasolar planets at the University of Cambridge who was not involved in the research, said several as yet poorly understood factors needed to be unpicked to make such estimates, including how life on Earth began and how many Earth-like planets considered habitable could truly support life.

Dr. Patricia Sanchez-Baracaldo, an expert on how Earth became habitable, from the University of Bristol, was more upbeat, despite emphasising that many developments were needed on Earth for conditions for complex life to exist, including photosynthesis. "But, yes, if we evolved in this planet, it is possible that intelligent life evolved in another part of the universe," she said.

Prof. Andrew Coates, of the Mullard Space Science Laboratory at University College London, said the assumptions made by Conselice and colleagues were reasonable, but the quest to find life was likely to take place closer to home for now.

"[The new estimate] is an interesting result, but one which it will be impossible to test using current techniques," he said. "In the meantime, research on whether we are alone in the universe will include visiting likely objects within our own solar system, for example with our Rosalind Franklin Exomars 2022 rover to Mars, and future missions to Europa, Enceladus and Titan [moons of Jupiter and Saturn]. It's a fascinating time in the search for life elsewhere."

> "[T]here is a fairly high likelihood that
> we are alone."

We're Probably Alone

Matt Williams

In the following viewpoint, Matt Williams reports on a study, published in 2018, that interprets the Fermi paradox in a new way. The result? Most likely, we're alone—at least in the galaxy. The truth is, there is so much scientific uncertainty about this that the conclusion is the equivalent of a shrug. The research team concluded that because there is so much uncertainty, we can say with greater confidence that humanity is most likely the only intelligent species. Matt Williams is a freelance writer, science fiction author, and curator of Universe Today's Guide to Space.

As you read, consider the following questions:

1. How, according to the scientists quoted in this article, is the Drake equation sensitive to bias?
2. Given the amount of uncertainty described in this study, how much do you think these new estimates change the discussion of the odds of life elsewhere?
3. How does the study of Earth's biology add to understanding of the likelihood that life arose elsewhere?

The Fermi Paradox remains a stumbling block when it comes to the search for extra-terrestrial intelligence (SETI). Named in honor of the famed physicist Enrico Fermi who first proposed it, this paradox addresses the apparent disparity between the expected probability that intelligent life is plentiful in the Universe, and the apparent lack of evidence of extra-terrestrial intelligence (ETI).

In the decades since Enrico Fermi first posed the question that encapsulates this paradox ("Where is everybody?"), scientists have attempted to explain this disparity one way or another. But in a new study conducted by three famed scholars from the Future of Humanity Institute (FHI) at Oxford University, the paradox is reevaluated in such a way that it makes it seem likely that humanity is alone in the observable Universe.

The study, titled "Dissolving the Fermi Paradox," recently appeared online. The study was jointly conducted by Anders Sandberg, a Research Fellow at the Future of Humanity Institute and a Martin Senior Fellow at Oxford University; Eric Drexler, the famed engineer who popularized the concept of nanotechnology; and Tod Ord, the famous Australian moral philosopher at Oxford University.

For the sake of their study, the team took a fresh look at the Drake Equation, the famous equation proposed by astronomer Dr. Frank Drake in the 1960s. Based on hypothetical values for a number of factors, this equation has traditionally been used to demonstrate that—even if the amount of life developing at any given site is small—the sheer multitude of possible sites should yield a large number of potentially observable civilizations.

This equation states that the number of civilizations (N) in our galaxy that we might be able to communicate with can be determined by multiplying the average rate of star formation in our galaxy (R^*), the fraction of those stars which have planets (fp), the number of planets that can actually support life (ne), the number of planets that will develop life (fl), the number of planets that will develop intelligent life (fi), the number of civilizations that would develop transmission technologies (fc), and the length of time that these civilizations would have to transmit their signals into space (L). Mathematically, this is expressed as:

$$N = R^* \text{ x } fp \text{ x } ne \text{ x } fl \text{ x } fi \text{ x } fc \text{ x } L$$

Dr. Sandberg is no stranger to the Fermi Paradox, nor is he shy about attempting to resolve it. In a previous study, titled "That is not dead which can eternal lie: the aestivation hypothesis for resolving Fermi's paradox," Sandberg and his associates proposed that the Fermi Paradox arises from the fact that ETIs are not dead, but currently in a state of hibernation—what they called "aestivation"—and awaiting better conditions in the Universe.

In a study conducted back in 2013, Sandberg and Stuart Armstrong (also a research associate with the FHI and one of the co-authors on this study) extended the Fermi Paradox to look beyond our own galaxy, addressing how more advanced civilizations would feasibly be able to launch colonization projects with relative ease (and even travel between galaxies without difficulty).

As Dr. Sandberg told Universe Today via email:

One can answer [the Fermi Paradox] by saying intelligence is very rare, but then it needs to be tremendously rare. Another possibility is that intelligence doesn't last very long, but it is enough that one civilization survives for it to become visible. Attempts at explaining it by having all intelligences acting in the same way (staying quiet, avoiding contact with us, transcending) fail since they require every individual belonging to every society in every civilization to behave in the same way, the strongest sociological claim ever. Claiming long-range settlement or communication are impossible requires assuming a surprisingly low technology ceiling. Whatever the answer is, it more or less has to be strange.

In this latest study, Sandberg, Drexler and Ord reconsider the parameters of the Drake Equation by incorporating models of chemical and genetic transitions on paths to the origin of life. From this, they show that there is a considerable amount of scientific uncertainties that span multiple orders of magnitude. Or as Dr. Sandberg explained it:

> Many parameters are very uncertain given current knowledge. While we have learned a lot more about the astrophysical ones since Drake and Sagan in the 1960s, we are still very uncertain about the probability of life and intelligence. When people discuss the equation it is not uncommon to hear them say something like: "this parameter is uncertain, but let's make a guess and remember that it is a guess," finally reaching a result that they admit is based on guesses. But this result will be stated as single number, and that anchors us to an *apparently* exact estimate—when it should have a proper uncertainty range. This often leads to overconfidence, and worse, the Drake equation is very sensitive to bias: if you are hopeful a small nudge upwards in several uncertain estimates will give a hopeful result, and if you are a pessimist you can easily get a low result.

As such, Sandberg, Drexler and Ord looked at the equation's parameters as uncertainty ranges. Instead of focusing on what value they might have, they looked at what the largest and smallest values they could have based on current knowledge. Whereas some values have become well constrained—such as the number of planets in our galaxy based on exoplanet studies and the number that exist within a star's habitable zone—others remain far more uncertain.

When they combined these uncertainties, rather than the guesswork that often go into the Fermi Paradox, the team got a distribution as a result. Naturally, this resulted in a broad spread due to the number of uncertainties involved. But as Dr. Sandberg explained, it did provide them with an estimate of the likelihood that humanity (given what we know) is alone in the galaxy:

> We found that even using the guesstimates in the literature (we took them and randomly combined the parameter estimates) one can have a situation where the mean number of civilizations

in the galaxy might be fairly high—say a hundred—and yet the probability that we are alone in the galaxy is 30%! The reason is that there is a very skew distribution of likelihood.

If we instead try to review the scientific knowledge, things get even more extreme. This is because the probability of getting life and intelligence on a planet has an *extreme* uncertainty given what we know—we cannot rule out that it happens nearly everywhere there is the right conditions, but we cannot rule out that it is astronomically rare. This leads to an even stronger uncertainty about the number of civilizations, drawing us to conclude that there is a fairly high likelihood that we are alone. However, we *also* conclude that we shouldn't be too surprised if we find intelligence!

In the end, the team's conclusions do not mean that humanity is alone in the Universe, or that the odds of finding evidence of extra-terrestrial civilizations (both past and present) is unlikely. Instead, it simply means that we can say with greater confidence—based on what we know—that humanity is most likely the only intelligent species in the Milky Way Galaxy at present.

And of course, this all comes down to the uncertainties we currently have to contend with when it comes to SETI and the Drake Equation. In that respect, the study conducted by Sandberg, Drexler and Ord is an indication that much more needs to be learned before we can attempt to determine just how likely ETI is out there.

"What we are not showing is that SETI is pointless—quite the opposite!" said Dr. Sandberg. "There is a tremendous level of uncertainty to reduce. The paper shows that astrobiology and SETI can play a big role in reducing the uncertainty about some of the parameters. Even terrestrial biology may give us important information about the probability of life emerging and the conditions leading to intelligence. Finally, one important conclusion we find is that lack of observed intelligence does not strongly make us conclude that intelligence doesn't last long: the stars are not foretelling our doom!"

So take heart, SETI enthusiasts! While the Drake Equation may not be something we can produce accurate values for anytime soon, the more we learn, the more refined the values will be. And remember, we only need to find intelligent life once in order for the Fermi Paradox to be resolved!

Periodical and Internet Sources Bibliography

The following articles have been selected to supplement the diverse views presented in this chapter.

Anil Ananthaswamy, "How Many Aliens Are in the Milky Way? Astronomers Turn to Statistics for Answers," *Scientific American*, July 16, 2020. https://www.scientificamerican.com/article /how-many-aliens-are-in-the-milky-way-astronomers-turn-to -statistics-for-answers/

Nadia Drake, "If Alien Life Exists in Our Solar System, It May Look Like This," *National Geographic*, November 11, 2019. https:// www.nationalgeographic.com/science/article/if-alien-life-exists -in-solar-system-may-look-like-this-aurora-hydrothermal-vent

Elizabeth Howell, "The Fermi Paradox: Where Are the Aliens?" Space.com, April 26, 2018. https://www.space.com/25325-fermi -paradox.html

Avery Hurt, "Life Elsewhere in the Universe: When Did We First Consider the Possibility?" *Discover*, March 22, 2021. https://www .discovermagazine.com/the-sciences/life-elsewhere-in-the -universe-when-did-we-first-consider-the-possibility

Elizabeth Kolbert, "Have We Already Been Visited by Aliens?" *New Yorker*, January 18, 2021. https://www.newyorker.com /magazine/2021/01/25/have-we-already-been-visited-by-aliens

Marina Koren, "If Aliens Are Out There, They're Way Out There," *The Atlantic*, May 25, 2021. https://www.theatlantic.com/science /archive/2021/05/ufos-aliens/618990/

Farhad Manjoo, "Aliens Must Be Out There. Why Aren't We Looking for Them?" *New York Times*, February 11, 2021. https://www .nytimes.com/2021/02/11/opinion/aliens-extraterrestrial-life .html

Dylan Matthews, "What If the Truth Isn't Out There? The Wishful Thinking Behind the Search for Alien Life," Vox, July 3, 2021. https://www.nytimes.com/2021/02/11/opinion/aliens -extraterrestrial-life.html

Elizabeth Rayne, "If We're Seeking Aliens, We Need to Look at Earth Through Their Eyes," SYFY Wire, April 5, 2021. https://www.syfy.com/syfywire/how-would-aliens-see-earth-if-they-exist

Dan Robitzski, "The Scientist Who Reevaluated the Drake Equation Still Thinks Alien Life Is Out There," Futurism, June 27, 2018. https://futurism.com/life-universe-scientist-drake-equation-study

Leonor Sierra, "Are We Alone in the Universe? Revisiting the Drake Equation," Exoplanet Exploration, May 19, 2016. https://exoplanets.nasa.gov/news/1350/are-we-alone-in-the-universe-revisiting-the-drake-equation/

Howard A. Smith, "Alone in the Universe," *American Scientist*, July/August 2011. https://www.americanscientist.org/article/alone-in-the-universe

Matt Williams, "What Are the Odds of Life Emerging on Another Planet?" Universe Today, June 3, 2020. https://www.universetoday.com/146308/what-are-the-odds-of-life-emerging-on-another-planet/

CHAPTER 2

Has Earth Already Been Visited by Extraterrestrial Life?

Chapter Preface

Chapter 1 viewpoints mentioned the possibility that Earth had already been visited by life forms from other planets. However, those viewpoints were mostly focused on the odds of alien life elsewhere in the universe. In this chapter, the viewpoints tackle the question "Are they here?" head-on. The question is not so much a theoretical "Were they here in the past" or "Are they under our feet in silicon form?" or even "If they're here, why haven't we heard from them?" but the far more immediate question of "Are they out there now?" In other words, the viewpoints in this chapter take on what you and I would call "the UFO Question."

However, you won't find any authors here claiming to have been abducted and probed by little green men, nor will you find "debunkers" explaining how ridiculous it is to believe in UFOs. Several of the authors in this chapter are themselves scientists, and the debate here is strictly from a scientific standpoint—one that evaluates the arguments and explores what little evidence there is. The first viewpoint is by an astronomer who, while admitting the possibility that intelligent aliens exist, says that scientists are waiting for better evidence. In a later viewpoint, we hear from a well-respected scientist who claims that an artifact of an alien civilization passed through our solar system as recently as 2017 and was observed by human scientists. The closing viewpoint points out several likely incorrect assumptions most of us (including many scientists) make when we talk about potential alien life. Getting rid of those assumptions opens the door for even more intriguing possibilities.

In between these two extremes, the authors of these viewpoints take varying approaches to the question. However, the days are long gone when even discussing the possibility that another advanced civilization might actually make contact could ruin a scientific career—though as you will see, most scientists are still being cautious about speculating on what they cannot prove.

> *"[Scientists] don't deny the existence of intelligent aliens. But they set a high bar for proof that we've been visited by creatures from another star system."*

Aliens May Already Be Out There, but Scientists Are Waiting for Better Evidence

Chris Impey

In the following viewpoint, Chris Impey begins by stating that, according to surveys, nearly half of Americans believe that aliens have visited Earth and that people in the United States are far more likely than people in other countries to claim to have seen UFOs. Scientists, on the other hand, are more skeptical, says Impey. While not ruling out the idea that alien ships may have visited Earth, they are waiting for better evidence. Chris Impey is professor of astronomy at the University of Arizona.

As you read, consider the following questions:

1. Why might dog walkers and smokers be more likely to see UFOs than other people?
2. How have UFOs become another religion?
3. Why doesn't the author think that belief in UFOs is crazy?

"I'm an Astronomer and I Think Aliens May Be Out There—but UFO Sightings Aren't Persuasive," by Chris Impey, The Conversation, December 4, 2020, https://theconversation.com/im-an-astronomer-and-i-think-aliens-may-be-out-there-but-ufo-sightings-arent-persuasive-150498. Licensed under CC BY-ND 4.0.

I f intelligent aliens visit the Earth, it would be one of the most profound events in human history.

Surveys show that nearly half of Americans believe that aliens have visited the Earth, either in the ancient past or recently. That percentage has been increasing. Belief in alien visitation is greater than belief that Bigfoot is a real creature, but less than belief that places can be haunted by spirits.

Scientists dismiss these beliefs as not representing real physical phenomena. They don't deny the existence of intelligent aliens. But they set a high bar for proof that we've been visited by creatures from another star system. As Carl Sagan said, "Extraordinary claims require extraordinary evidence."

I'm a professor of astronomy who has written extensively on the search for life in the universe. I also teach a free online class on astrobiology. Full disclosure: I have not personally seen a UFO.

Unidentified Flying Objects

UFO means unidentified flying object. Nothing more, nothing less.

There's a long history of UFO sightings. Air Force studies of UFOs have been going on since the 1940s. In the United States, "ground zero" for UFOs occurred in 1947 in Roswell, New Mexico. The fact that the Roswell incident was soon explained as the crash landing of a military high-altitude balloon didn't stem a tide of new sightings. The majority of UFOs appear to people in the United States. It's curious that Asia and Africa have so few sightings despite their large populations, and even more surprising that the sightings stop at the Canadian and Mexican borders.

Most UFOs have mundane explanations. Over half can be attributed to meteors, fireballs and the planet Venus. Such bright objects are familiar to astronomers but are often not recognized by members of the public. Reports of visits from UFOs inexplicably peaked about six years ago.

Many people who say they have seen UFOs are either dog walkers or smokers. Why? Because they're outside the most.

Sightings concentrate in evening hours, particularly on Fridays, when many people are relaxing with one or more drinks.

A few people, like former NASA employee James Oberg, have the fortitude to track down and find conventional explanations for decades of UFO sightings. Most astronomers find the hypothesis of alien visits implausible, so they concentrate their energy on the exciting scientific search for life beyond the Earth.

Are We Alone?

While UFOs continue to swirl in the popular culture, scientists are trying to answer the big question that is raised by UFOs: Are we alone?

Astronomers have discovered over 4,000 exoplanets, or planets orbiting other stars, a number that doubles every two years. Some of these exoplanets are considered habitable, since they are close to the Earth's mass and at the right distance from their stars to have water on their surfaces. The nearest of these habitable planets are less than 20 light years away, in our cosmic "back yard." Extrapolating from these results leads to a projection of 300 million habitable worlds in our galaxy. Each of these Earth-like planets is a potential biological experiment, and there have been billions of years since they formed for life to develop and for intelligence and technology to emerge.

Astronomers are very confident there is life beyond the Earth. As astronomer and ace exoplanet-hunter Geoff Marcy, puts it, "The universe is apparently bulging at the seams with the ingredients of biology." There are many steps in the progression from Earths with suitable conditions for life to intelligent aliens hopping from star to star. Astronomers use the Drake Equation to estimate the number of technological alien civilizations in our galaxy. There are many uncertainties in the Drake Equation, but interpreting it in the light of recent exoplanet discoveries makes it very unlikely that we are the only, or the first, advanced civilization.

This confidence has fueled an active search for intelligent life, which has been unsuccessful so far. So researchers have recast the question "Are we alone?" to "Where are they?"

The absence of evidence for intelligent aliens is called the Fermi Paradox. Even if intelligent aliens do exist, there are a number of reasons why we might not have found them and they might not have found us. Scientists do not discount the idea of aliens. But they aren't convinced by the evidence to date because it is unreliable, or because there are so many other more mundane explanations.

Modern Myth and Religion

UFOs are part of the landscape of conspiracy theories, including accounts of abduction by aliens and crop circles created by aliens. I remain skeptical that intelligent beings with vastly superior technology would travel trillion of miles just to press down our wheat.

It's useful to consider UFOs as a cultural phenomenon. Diana Pasulka, a professor at the University of North Carolina, notes that myths and religions are both means for dealing with unimaginable experiences. To my mind, UFOs have become a kind of new American religion.

So no, I don't think belief in UFOs is crazy, because some flying objects are unidentified, and the existence of intelligent aliens is scientifically plausible.

But a study of young adults did find that UFO belief is associated with schizotypal personality, a tendency toward social anxiety, paranoid ideas and transient psychosis. If you believe in UFOs, you might look at what other unconventional beliefs you have.

I'm not signing on to the UFO "religion," so call me an agnostic. I recall the aphorism popularized by Carl Sagan, "It pays to keep an open mind, but not so open your brains fall out."

> *"Claims like this, especially from experienced scientists are disliked by the scientific community for many reasons."*

Extraordinary Claims from Experienced Scientists Are Disliked by the Scientific Community

Simon Goodwin

Scientists may be willing to admit the possibility that life exists elsewhere in the universe, but few are willing to go so far as to say they've already been here. In the following viewpoint, Simon Goodwin digs into the controversy ignited by Avi Loeb's paper suggesting that 'Oumuamua, an interstellar object passing by in 2017, might be of alien origin. Simon Goodwin is professor of theoretical astrophysics at the University of Sheffield in the UK.

As you read, consider the following questions:

1. How many comets and asteroids are in the Milky Way?
2. What is a light sail?
3. What is Occam's razor?

"Has Earth Been Visited by an Alien Spaceship? Harvard Professor Avi Loeb vs Everybody Else," by Simon Goodwin, The Conversation, February 18, 2021, https://theconversation .com/has-earth-been-visited-by-an-alien-spaceship-harvard-professor-avi-loeb-vs -everybody-else-155509. Licensed under CC BY-ND 4.0.

A highly unusual object was spotted travelling through the solar system in 2017. Given a Hawaiian name, 'Oumuamua, it was small and elongated—a few hundred metres by a few tens of meters, travelling at a speed fast enough to escape the Sun's gravity and move into interstellar space.

I was at a meeting when the discovery of 'Oumuamua was announced, and a friend immediately said to me, "So how long before somebody claims it's a spaceship?" It seems that whenever astronomers discover anything unusual, somebody claims it must be aliens.

Nearly all scientists believe that 'Oumuamua probably originated from outside the solar system. It is an asteroid—or comet-like object that has left another star and travelled through interstellar space—we saw it as it zipped by us. But not everyone agrees. Avi Loeb, a Harvard professor of astronomy, suggested in a recent book that it is indeed an alien spaceship. But how feasible is this? And how come most scientists disagree with the claim?

Researchers estimate that the Milky Way should contain around 100 million billion billion comets and asteroids ejected from other planetary systems, and that one of these should pass through our solar system every year or so. So it makes sense that 'Oumuamua could be one of these. We spotted another last year—"Borisov"—which suggests they are as common as we predict.

What made 'Oumuamua particularly interesting was that it didn't follow the orbit you would expect—its trajectory shows it has some extra "non-gravitational force" acting on it. This is not too unusual. The pressure of solar radiation or gas or particles driven out as an object warms up close to the Sun can give extra force, and we see this with comets all the time.

Experts on comets and the solar system have explored various explanations for this. Given this was a small, dark object passing us very quickly before disappearing, the images we were able to get weren't wonderful, and so it is difficult to be sure.

LIFE ON VENUS? NASA'S LOOKING INTO IT

[On Monday, scientific research was published showing phosphine, a possible signature of life, present in the atmosphere of Venus. The following is NASA's announcement related to this discovery. –ed.]

A paper about chemistry on Venus was recently published in *Nature Astronomy*. NASA was not involved in the research and cannot comment directly on the findings; however, we trust in the scientific peer review process and look forward to the robust discussion that will follow its publication.

NASA has an extensive astrobiology program that searches for life in many different ways across the solar system and beyond. Over the past two decades, we've made new discoveries that collectively imply a significant increase of the likelihood to find life elsewhere.

As with an increasing number of planetary bodies, Venus is proving to be an exciting place of discovery, though it had not been a significant part of the search for life because of its extreme temperatures, atmospheric composition and other factors. Two of the next four candidate missions for NASA's Discovery Program are focused on Venus, as is Europe's EnVision mission, in which NASA is a partner. Venus also is a planetary destination we can reach with smaller missions.

Astrobiology is the study of the origin, evolution, and distribution of life in the universe.

"NASA Issues Announcement Regarding Claims of Possible Signs of Life on Venus," by NASA, Scitechdaily, September 16, 2020.

Loeb, however, believes that 'Oumuamua is an alien spaceship, powered by a "light sail"—a method for propelling a spacecraft using radiation pressure exerted by the Sun on huge mirrors. He argues the non-gravitational acceleration is a sign of "deliberate" manoeuvring. This argument seems largely to be based on the fact that 'Oumuamua lacks a fuzzy envelope ("coma") and a comet-like tail, which are usual signatures of comets undergoing non-gravitational acceleration (although jets from particular spots cannot be ruled out).

Sanity Checks

He may or may not be right, and there is no way of proving or disproving this idea. But claims like this, especially from experienced scientists are disliked by the scientific community for many reasons.

If we decide that anything slightly odd that we don't understand completely in astronomy could be aliens, then we have a lot of potential evidence for aliens—there is an awful lot we don't understand. To stop ourselves jumping to weird and wonderful conclusions every time we come across something strange, science has several sanity checks.

One is Occam's razor, which tells us to look for the simplest solutions that raise the fewest new questions. Is this a natural object of the type that we suspect to be extremely common in the Milky Way, or is it aliens? Aliens raise a whole set of supplementary questions (who, why, from where?), which means Occam's razor tells us to reject it, at least until all simpler explanations are exhausted.

Another sanity check is the general rule that "extraordinary claims require extraordinary evidence." A not quite completely understood acceleration is not extraordinary evidence, as there are many plausible explanations for it.

Yet another check is the often sluggish but usually reliable peer-review system, in which scientists publish their findings in scientific journals where their claims can be assessed and critiqued by experts in their field.

Alien Research

This doesn't mean that we shouldn't look for aliens. A lot of time and money is being devoted to researching them. For astronomers who are interested in the proper science of aliens, there is "astrobiology"—the science of looking for life outside Earth based on signs of biological activity. On February 18, NASA's *Perseverance* rover will land on Mars and look for molecules which may include such signatures, for example. Other interesting targets are the moons of Jupiter and Saturn.

In the next five years, we will also have the technology to search for alien life on planets around other stars (exoplanets). Both the James Webb Space Telescope (due to launch in 2021) and the European Extremely Large Telescope (due for first light in 2025) will analyse exoplanet atmospheres in detail, searching for signs of life. For example, the oxygen in the Earth's atmosphere is there because life constantly produces it. Meanwhile, the Search for Extraterrestrial Intelligence (Seti) initiative has been scanning the skies with radio telescopes for decades in search of messages from intelligent aliens.

Signs of alien life would be an amazing discovery. But when we do find such evidence, we want to be sure it is good. To be as sure as we can be, we need to present our arguments to other experts in the field to examine and critique, follow the scientific method which, in its slow and plodding way, gets us there in the end.

This would give us much more reliable evidence than claims from somebody with a book to sell. It is quite possible, in the next five to ten years, that somebody will announce that they have found good evidence for alien life. But rest assured this isn't it.

> *"Ongoing efforts by military insiders to influence policy, without proper context or analysis, reflect a worrying breakdown of the intelligence cycle."*

The UFO Craze May Be a Problem of Intelligence Failings

Kyle Cunliffe

The government used to call them UFOs. Now they're called UAP (unidentified aerial phenomena). In the following viewpoint, Kyle Cunliffe discusses a recent report by US intelligence on recent sightings by military personnel of apparently flying objects that can't be identified. The report, writes the author, is inconclusive. And that, he argues, raises as many questions about US intelligence as it does about UFOs. Kyle Cunliffe is a lecturer at the University of Salford in Manchester, England.

As you read, consider the following questions:

1. Why are analysts unable to explain all but one sighting?
2. What were some of the problems and contradictions in the report of the sightings on the USS *Nimitz*?
3. What motives may be behind the leak of UAP sightings to the press?

"The Truth Is Still Out There: Why the Current UFO Craze May Be a Problem of Intelligence Failings," by Kyle Cunliffe, The Conversation, June 29, 2021, https://theconversation.com/the-truth-is-still-out-there-why-the-current-ufo-craze-may-be-a-problem-of-intelligence-failings-163185. Licensed under CC BY-ND 4.0.

It's safe to say that UFOs, now branded UAPs, are back. In recent years, concerns have grown that supposed physics-defying craft are penetrating US airspace. This could represent a technological breakthrough by foreign competitors or something else entirely. But many people will no doubt have found the recent release of the Pentagon's highly anticipated UAP (unidentified aerial phenomena) report to be underwhelming.

Its results are inconclusive, despite the fact that it is the alleged weight of the data that led Congress to request the report in the first place. This raises serious questions as to how the intelligence process became so muddied, and why UFOs have rocketed up Washington's agenda.

While it puts many hypotheses forward, the report concedes that analysts cannot explain at least 143 out of 144 reported sightings. The problem, as they acknowledge, is that they lack the data to draw firm conclusions. The issue is not simply about whether the extraordinary things that have been reported belong to Russia, China or the Klingons, but more about whether anything extraordinary is even happening at all.

To an extent, this is unsurprising. In practically every UAP incident reported, nobody can agree whether something extraordinary—a physics-bending craft, for example—was actually witnessed. Sceptics argue that factors such as misreporting, technical and human error, or optical illusions, can explain much of what is happening in the skies.

Nimitz Encounter

This is personified in the 2004 *Nimitz* encounter where two pilots spotted a white object shaped like a "Tic Tac." The erratic craft reportedly responded to the pilots' movements, before disappearing in a blink of an eye. It reappeared sometime later, where a third pilot recorded footage that would eventually make its way to the *New York Times* in 2017.

The encounter was allegedly investigated by the Pentagon's AATIP (Advanced Aerospace Threat Identification Program), which has since been renamed the UAP Task Force—the body now responsible for the UAP report. And it gained traction thanks to the openness of one of its star witnesses, pilot Commander David Fravor, who told ABC News that the Tic Tac seemed "not from this world."

The case, however, seems riddled with issues of reporting and human testimony. Fravor has dismissed claims by other crew from the *Nimitz* carrier group, including allegations that mysterious officials requisitioned crucial data. And the other pilot at the time of the first encounter, Alex Dietrich, claimed that her visual on the Tic Tac lasted around ten seconds—a stark comparison to Fravor's claim of five minutes.

The point is that memory and misperception affect even the best-trained pilots. Notable sceptic Mick West argues that optical illusions can explain away much of the pilot and video testimony, and the report itself concedes that "observer misperception" cannot be ruled out in some sightings.

The *Nimitz* case, as with other UAP incidents, was supported by radar and sensor data—but this is yet to be revealed to the public. And it bears consideration that even the most expensive technical systems are not infallible. As the report acknowledges, cases where UAPs exhibited "unusual flight characteristics" may also be the result of sensor errors or "spoofing"—a known technical countermeasure that tricks radar systems into displaying inaccurate information.

Need to Know

These challenges filter down to analysts, who face an overwhelming task. In fact, UAP analysts are relying on intelligence collection systems to answer what is essentially a scientific problem. As the report notes, US military sensors are "designed to fulfill a specific mission," and are not "generally suited for identifying UAP."

It's more likely that understanding the problem will require a myriad of technical instruments supported by scientific collaboration and peer review, which runs to the contrary of intelligence's "need to know." If there are any extraordinary answers to be had, they are more likely to come from the recent involvement of NASA, than the closed-door world of the UAP task force.

What's more, faced with limited data, analysts are vulnerable to their own cognitive biases. AATIP was originally contracted to a company whose founder, Robert Bigelow, is well known for his paranormal enthusiasm. And AATIP's former director, Luis Elizondo, continues to push the narrative that UAPs are real craft and possibly of non-human origin.

And then there's the issue of inflation. The official, Christopher Mellon, who first set events in motion by leaking the 2017 footage to the *New York Times*, admits that he and Elizondo wanted to put UAPs on the "national security agenda." Policymakers should be led by refined intelligence assessments, not the personal hunches of analysts and officials whose opinions are shaped by mediocre data.

Bomber Gap

Indeed, current events are not dissimilar to the cold war's "bomber gap," when Air Force analysts vastly inflated Soviet nuclear bomber estimates to secure greater Congressional funding. As a result of Elizondo and Mellon's efforts, UAPs are now on the agenda, whether they exist or not. Even the report calls for "analytical, collection, and resource investment."

But as Congress demands further investigation, it should also demand greater accountability. Authenticated (albeit mundane) military footage of UAPs continues to be leaked to UFO film makers. These ongoing efforts by military insiders to influence policy, without proper context or analysis, reflect a worrying breakdown of the intelligence cycle.

Finally, there's the issue of politicisation. AATIP was originally established by the former Senate majority leader Harry Reid under advice from his close friend Bigelow. Reid's enthusiasm for UFOs is well documented, but it suggests that the process was muddied from the start. If the UAP Task Force is expanded, a healthy distance will need to be maintained between policymakers and the people who draw up their assessments.

As it stands, the UAP issue seems like a microcosm of everything that can go wrong with intelligence. If the UAP report suggests anything, it's that pilots are struggling to make sense of increasingly noisy skies, that military sensors cannot always be relied upon, and that the Pentagon's analysts are out of their depth.

It also shows that unless the Department of Defense obtains clear evidence of an undeniable craft operating in undeniably extraordinary ways, Congress, and the public, should remain sceptical of UAP proponents.

> *"Although most of the UAP described in our dataset probably remain unidentified due to limited data or challenges to collection processing or analysis, we may require additional scientific knowledge to successfully collect on, analyze and characterize some of them."*

The Office of the Director of National Intelligence's Preliminary Assessment on Unidentified Aerial Phenomena

Office of the Director of National Intelligence

In the following viewpoint, the Office of the Director of National Intelligence (ODNI) reports on its investigation into UAP. The federally mandated investigation was part of the Senate's Intelligence Authorization Act of 2021, in large response to reports of UFOs by US Navy pilots, with one aim being the analysis of security threat to the US. While it's ultimately inconclusive, the report is considered the first step into finding whether the aerial threats pose serious concerns to our national safety. The Office of the Director of National Intelligence is a senior-level agency that provides oversight to the Intelligence Community.

"Preliminary Assessment: Unidentified Aerial Phenomena," Office of the Director of National Intelligence, June 25, 2021.

As you read, consider the following questions:

1. What do the authors of this report say are the reasons they are unable to give more definitive answers?
2. According to this report, what sorts of risk and challenges do UAP present?
3. What does the report say is needed in order to provide better explanations for the mysterious phenomena?

Scope and Assumptions

Scope

This preliminary report is provided by the Office of the Director of National Intelligence (ODNI) in response to the provision in Senate Report 116-233, accompanying the Intelligence Authorization Act (IAA) for Fiscal Year 2021, that the DNI, in consultation with the Secretary of Defense (SECDEF), is to submit an intelligence assessment of the threat posed by unidentified aerial phenomena (UAP) and the progress the Department of Defense Unidentified Aerial Phenomena Task Force (UAPTF) has made in understanding this threat.

This report provides an overview for policymakers of the challenges associated with characterizing the potential threat posed by UAP while also providing a means to develop relevant processes, policies, technologies, and training for the US military and other US Government (USG) personnel if and when they encounter UAP, so as to enhance the Intelligence Community's (IC) ability to understand the threat. The Director, UAPTF, is the accountable official for ensuring the timely collection and consolidation of data on UAP. The dataset described in this report is currently limited primarily to US Government reporting of incidents occurring from November 2004 to March 2021. Data continues to be collected and analyzed.

ODNI prepared this report for the Congressional Intelligence and Armed Services Committees. UAPTF and the ODNI National Intelligence Manager for Aviation drafted this report, with input from USD(I&S), DIA, FBI, NRO, NGA, NSA, Air Force, Army, Navy, Navy/ONI, DARPA, FAA, NOAA, NGA, ODNI/ NIM-Emerging and Disruptive Technology, ODNI/National Counterintelligence and Security Center, and ODNI/National Intelligence Council.

Assumptions

Various forms of sensors that register UAP generally operate correctly and capture enough real data to allow initial assessments, but some UAP may be attributable to sensor anomalies.

Executive Summary

The limited amount of high-quality reporting on unidentified aerial phenomena (UAP) hampers our ability to draw firm conclusions about the nature or intent of UAP. The Unidentified Aerial Phenomena Task Force (UAPTF) considered a range of information on UAP described in US military and IC (Intelligence Community) reporting, but because the reporting lacked sufficient specificity, ultimately recognized that a unique, tailored reporting process was required to provide sufficient data for analysis of UAP events.

- As a result, the UAPTF concentrated its review on reports that occurred between 2004 and 2021, the majority of which are a result of this new tailored process to better capture UAP events through formalized reporting.
- Most of the UAP reported probably do represent physical objects given that a majority of UAP were registered across multiple sensors, to include radar, infrared, electro-optical, weapon seekers, and visual observation.

In a limited number of incidents, UAP reportedly appeared to exhibit unusual flight characteristics. These observations could be the result of sensor errors, spoofing, or observer misperception and require additional rigorous analysis.

There are probably multiple types of UAP requiring different explanations based on the range of appearances and behaviors described in the available reporting. Our analysis of the data supports the construct that if and when individual UAP incidents are resolved they will fall into one of five potential explanatory categories: airborne clutter, natural atmospheric phenomena, USG or US industry developmental programs, foreign adversary systems, and a catchall "other" bin.

UAP clearly pose a safety of flight issue and may pose a challenge to US national security. Safety concerns primarily center on aviators contending with an increasingly cluttered air domain. UAP would also represent a national security challenge if they are foreign adversary collection platforms or provide evidence a potential adversary has developed either a breakthrough or disruptive technology.

Consistent consolidation of reports from across the federal government, standardized reporting, increased collection and analysis, and a streamlined process for screening all such reports against a broad range of relevant USG data will allow for a more sophisticated analysis of UAP that is likely to deepen our understanding. Some of these steps are resource-intensive and would require additional investment.

Available Reporting Largely Inconclusive
Limited Data Leaves Most UAP Unexplained...

Limited data and inconsistency in reporting are key challenges to evaluating UAP. No standardized reporting mechanism existed until the Navy established one in March 2019. The Air Force subsequently adopted that mechanism in November 2020, but it remains limited to USG reporting. The UAPTF regularly heard anecdotally during its research about other observations that occurred but which were never captured in formal or informal reporting by those observers.

After carefully considering this information, the UAPTF focused on reports that involved UAP largely witnessed firsthand by military aviators and that were collected from systems we considered to be reliable. These reports describe incidents that occurred between 2004 and 2021, with the majority coming in the last two years as the new reporting mechanism became better known to the military aviation community. We were able to identify one reported UAP with high confidence. In that case, we identified the object as a large, deflating balloon. The others remain unexplained.

- **144** reports originated from USG sources. Of these, **80** reports involved observation with multiple sensors.
- Most reports described UAP as objects that interrupted pre-planned training or other military activity.

UAP Collection Challenges

Sociocultural stigmas and sensor limitations remain obstacles to collecting data on UAP. Although some technical challenges — such as how to appropriately filter out radar clutter to ensure safety of flight for military and civilian aircraft — are longstanding in the aviation community, others are unique to the UAP problem set.

- Narratives from aviators in the operational community and analysts from the military and IC describe disparagement associated with observing UAP, reporting it, or attempting to discuss it with colleagues. Although the effects of these stigmas have lessened as senior members of the scientific, policy, military, and intelligence communities engage on the topic seriously in public, reputational risk may keep many observers silent, complicating scientific pursuit of the topic.
- The sensors mounted on US military platforms are typically designed to fulfill specific missions. As a result, those sensors are not generally suited for identifying UAP.
- Sensor vantage points and the numbers of sensors concurrently observing an object play substantial roles in distinguishing UAP from known objects and determining whether a UAP demonstrates breakthrough aerospace capabilities. Optical sensors have the benefit of providing some insight into relative size, shape, and structure. Radiofrequency sensors provide more accurate velocity and range information.

But Some Potential Patterns Do Emerge

Although there was wide variability in the reports and the dataset is currently too limited to allow for detailed trend or pattern analysis, there was some clustering of UAP observations regarding shape, size, and, particularly, propulsion. UAP sightings also tended to cluster around US training and testing grounds, but we assess that this may result from a collection bias as a result of focused attention, greater numbers of latest-generation sensors operating in those areas, unit expectations, and guidance to report anomalies.

And a Handful of UAP Appear to Demonstrate Advanced Technology

In **18** incidents, described in **21** reports, observers reported unusual UAP movement patterns or flight characteristics.

Some UAP appeared to remain stationary in winds aloft, move against the wind, maneuver abruptly, or move at considerable speed, without discernable means of propulsion. In a small number of cases, military aircraft systems processed radio frequency (RF) energy associated with UAP sightings.

The UAPTF holds a small amount of data that appear to show UAP demonstrating acceleration or a degree of signature management. Additional rigorous analysis are necessary by multiple teams or groups of technical experts to determine the nature and validity of these data. We are conducting further analysis to determine if breakthrough technologies were demonstrated.

UAP Probably Lack a Single Explanation

The UAP documented in this limited dataset demonstrate an array of aerial behaviors, reinforcing the possibility there are multiple types of UAP requiring different explanations. Our analysis of the data supports the construct that if and when individual UAP incidents are resolved they will fall into one of five potential explanatory categories: airborne clutter, natural atmospheric phenomena, USG or industry developmental programs, foreign adversary systems, and a catchall "other" bin. With the exception of the one instance where we determined with high confidence that the reported UAP was airborne clutter, specifically a deflating balloon, we currently lack sufficient information in our dataset to attribute incidents to specific explanations.

Airborne Clutter: These objects include birds, balloons, recreational unmanned aerial vehicles (UAV), or airborne debris like plastic bags that muddle a scene and affect an operator's ability to identify true targets, such as enemy aircraft.

Natural Atmospheric Phenomena: Natural atmospheric phenomena includes ice crystals, moisture, and thermal fluctuations that may register on some infrared and radar systems.

USG or Industry Developmental Programs: Some UAP observations could be attributable to developments and classified programs by US entities. We were unable to confirm, however, that these systems accounted for any of the UAP reports we collected.

Foreign Adversary Systems: Some UAP may be technologies deployed by China, Russia, another nation, or a non-governmental entity.

Other: Although most of the UAP described in our dataset probably remain unidentified due to limited data or challenges to collection processing or analysis, we may require additional scientific knowledge to successfully collect on, analyze and characterize some of them. We would group such objects in this category pending scientific advances that allowed us to better understand them. The UAPTF intends to focus additional analysis on the small number of cases where a UAP appeared to display unusual flight characteristics or signature management.

UAP Threaten Flight Safety and, Possibly, National Security

UAP pose a hazard to safety of flight and could pose a broader danger if some instances represent sophisticated collection against US military activities by a foreign government or demonstrate a breakthrough aerospace technology by a potential adversary.

Ongoing Airspace Concerns

When aviators encounter safety hazards, they are required to report these concerns. Depending on the location, volume, and behavior of hazards during incursions on ranges, pilots may cease their tests and/or training and land their aircraft, which has a deterrent effect on reporting.

The UAPTF has 11 reports of documented instances in which pilots reported near misses with a UAP.

Potential National Security Challenges

We currently lack data to indicate any UAP are part of a foreign collection program or indicative of a major technological advancement by a potential adversary. We continue to monitor for evidence of such programs given the counter intelligence challenge they would pose, particularly as some UAP have been detected near military facilities or by aircraft carrying the USG's most advanced sensor systems.

Explaining UAP Will Require Analytic, Collection and Resource Investment

Standardize the Reporting, Consolidate the Data, and Deepen the Analysis

In line with the provisions of Senate Report 116-233, accompanying the IAA for FY 2021, the UAPTF's long-term goal is to widen the scope of its work to include additional UAP events documented by a broader swath of USG personnel and technical systems in its analysis. As the dataset increases, the UAPTF's ability to employ data analytics to detect trends will also improve. The initial focus will be to employ artificial intelligence/machine learning algorithms to cluster and recognize similarities and patterns in features of the data points. As the database accumulates information from known aerial objects such as weather balloons, high-altitude or super-pressure balloons, and wildlife, machine learning can add efficiency by pre-assessing UAP reports to see if those records match similar events already in the database.

The UAPTF has begun to develop interagency analytical and processing workflows to ensure both collection and analysis will be well informed and coordinated.

The majority of UAP data is from US Navy reporting, but efforts are underway to standardize incident reporting across US military services and other government agencies to ensure all relevant data is captured with respect to particular incidents and any US activities that might be relevant. The UAPTF is currently working to acquire additional reporting, including from the US Air Force (USAF), and has begun receiving data from the Federal Aviation Administration (FAA).

Although USAF data collection has been limited historically the USAF began a six-month pilot program in November 2020 to collect in the most likely areas to encounter UAP and is evaluating how to normalize future collection, reporting, and analysis across the entire Air Force.

The FAA captures data related to UAP during the normal course of managing air traffic operations. The FAA generally ingests this data when pilots and other airspace users report unusual or unexpected events to the FAA's Air Traffic Organization.

In addition, the FAA continuously monitors its systems for anomalies, generating additional information that may be of use to the UAPTF. The FAA is able to isolate data of interest to the UAPTF and make it available. The FAA has a robust and effective outreach program that can help the UAPTF reach members of the aviation community to highlight the importance of reporting UAP.

Expand Collection

The UAPTF is looking for novel ways to increase collection of UAP cluster areas when US forces are not present as a way to baseline "standard" UAP activity and mitigate the collection bias in the dataset. One proposal is to use advanced algorithms to search historical data captured and stored by radars. The UAPTF also plans to update its current interagency UAP collection strategy in order bring to bear relevant collection platforms and methods from the DoD and the IC.

Increase Investment in Research and Development

The UAPTF has indicated that additional funding for research and development could further the future study of the topics laid out in this report. Such investments should be guided by a UAP Collection Strategy, UAP R&D Technical Roadmap, and a UAP Program Plan.

Appendix A—Definition of Key Terms

This report and UAPTF databases use the following defining terms:

Unidentified Aerial Phenomena (UAP): Airborne objects not immediately identifiable. The acronym UAP represents the broadest category of airborne objects reviewed for analysis.

UAP Event: A holistic description of an occurrence during which a pilot or aircrew witnessed (or detected) a UAP.

UAP Incident: A specific part of the event.

UAP Report: Documentation of a UAP event, to include verified chains of custody and basic information such as the time, date, location, and description of the UAP. UAP reports include Range Fouler reports and other reporting. [US Navy aviators define a "range fouler" as an activity or object that interrupts pre-planned training or other military activity in a military operating area or restricted airspace.]

Appendix B—Senate Report Accompanying the Intelligence Authorization Act for Fiscal Year 2021

Senate Report 116-233, accompanying the Intelligence Authorization Act for Fiscal Year 2021, provides that the DNI, in consultation with the SECDEF and other relevant heads of USG Agencies, is to submit an intelligence assessment of the threat posed by UAP and the progress the UAPTF has made to understand this threat.

The Senate Report specifically requested that the report include:

1. A detailed analysis of UAP data and intelligence reporting collected or held by the Office of Naval Intelligence, including data and intelligence reporting held by the UAPTF;
2. A detailed analysis of unidentified phenomena data collected by:
   a. Geospatial Intelligence;
   b. Signals Intelligence;
   c. Human Intelligence; and
   d. Measurement and Signatures Intelligence
3. A detailed analysis of data of the Federal Bureau of Investigation, which was derived from investigations of intrusions of UAP data over restricted US airspace;
4. A detailed description of an interagency process for ensuring timely data collection and centralized analysis of all UAP reporting for the Federal Government, regardless of which service or agency acquired the information;
5. Identification of an official accountable for the process described in paragraph 4;
6. Identification of potential aerospace or other threats posed by the UAP to national security, and an assessment of whether this UAP activity may be attributed to one or more foreign adversaries;

7. Identification of any incidents or patterns that indicate a potential adversary, have achieved breakthrough aerospace capabilities that could put US strategic or conventional forces at risk; and

8. Recommendations regarding increased collection of data, enhanced research and development, additional funding, and other resources.

Periodical and Internet Sources Bibliography

The following articles have been selected to supplement the diverse views presented in this chapter.

Kate Dorsch, "UFOs Were Born Among America's Cold War Fears," *FP*, June 6, 2021. https://foreignpolicy.com/2021/06/06/ufos -space-cold-war-pentagon-military-aliens/

Lillian Steenblik Hwang, "50 Years Ago, UFO Sightings in the United States Went Bust," *Science News*, June 21, 2021. https://www .sciencenews.org/article/50-years-ago-ufo-sightings-united -states-went-bust

Chris Impey, "US Intelligence Report on UFOs: No Aliens, but Government Transparency and Desire for Better Data Might Bring Science to the UFO World," The Conversation, June 30, 2021. https://theconversation.com/us-intelligence-report-on -ufos-no-aliens-but-government-transparency-and-desire-for -better-data-might-bring-science-to-the-ufo-world-163059

Miriam Kramer, "Why UFO Sightings Are Here to Stay," Axios, June 26, 2021. https://www.axios.com/ufo-report-pentagon-damage -done-f7dfb6c0-052c-40f1-836b-aa30f58cf532.html

Rafi Letzter, "Mystery of Interstellar Visitor 'Oumuamua Gets Trickier," *Scientific American*, August 19, 2020. https://www .scientificamerican.com/article/mystery-of-interstellar-visitor -oumuamua-gets-trickier/

Gideon Lewis-Kraus, "How the Pentagon Started Taking U.F.O.s Seriously," *New Yorker*, April 30, 2021. https://www.newyorker .com/magazine/2021/05/10/how-the-pentagon-started-taking- ufos-seriously

Avi Loeb, "Could There Be a Link Between Interstellar Visitor 'Oumuamua and Unidentified Aerial Phenomena?" Space.com, June 29, 2021. https://www.space.com/oumuamua-unidentified -aerial-phenomena

Sarah Marquart, "Yes, a NASA Scientist Said Aliens May Have Visited Earth. But There's Some Nuance," Futurism, December 5, 2018. https://futurism.com/nasa-scientist-aliens-visited-earth-nuance

Albert McKeon, "NOW Do Aliens Exist? Scientists Wonder if Extraterrestrial Life Has Visited Earth," Northrop Grumman,

June 15, 2020. https://now.northropgrumman.com/scientists
-wonder-if-extraterrestrial-life-has-visited-earth/

Dennis Overbye, "Why Oumuamua, the Interstellar Visitor, Looks Eerily Familiar," *New York Times*, March 23, 2021. https://www
.nytimes.com/2021/03/23/science/astronomy-oumuamua-comet
.html

Sarah Scoles, "The UFO Trap," *The Atlantic*, June 25, 2021. https://
www.theatlantic.com/technology/archive/2021/06/ufo-report
-uap-director-national-intellegence/619293/

Wired, "How UFO Sightings Became an American Obsession," March 3, 2020. https://www.wired.com/story/how-ufo-sightings
-became-an-american-obsession/

OPPOSING
VIEWPOINTS®
SERIES

CHAPTER 3

Is It Foolhardy to Send Out Signals Letting Potential Enemies Know We Are Here?

Chapter Preface

Humans have long been fascinated by the idea that we are not alone, that there are other life forms similar to ours elsewhere in the universe, maybe even in our own galaxy. However, only recently have we had the technology to reach out to them. We've done this in a variety of ways. We've sent radio signals into space. We've launched spacecraft containing messages and artifacts from Earth. So far, we've been met with silence (or as one writer points out, if there has been a response, we've not been able to recognize it as such).

As exciting as these efforts are, not everyone thinks this is such a good idea. After all, if there are intelligent life forms out there, we have no idea if they are peaceful or bent on destroying anyone else they encounter. Contacting another life form might be the height of our technology. It might also be the last thing we do. The authors in this chapter discuss the pros and cons of reaching out to whoever might be out there and thus alerting them to our presence before we know if they have benign or evil intentions toward us.

The first viewpoint sets up the debate, covering a conference at which the subject was discussed, and including the points of view of people on both sides of the issue. Then we hear from a variety of authors, many who think it's too late to worry about it; if they're out there, they already know we're here, thanks to our bright and noisy planet. Others point out that trying to hide is actually riskier than reaching out, because trying to silence (and darken) ourselves would hamper important technologies.

Others, however, are seriously concerned about the risk of identifying ourselves to potentially aggressive civilizations with the power to do us harm. One viewpoint suggests that quantum communications might one day offer a safer way to communicate. The last viewpoint sums up the arguments and ends on a slightly humorous and less alarming note.

> *"Most agree that the search for intelligent life is a smart idea . . . But some believe that sending signals and information into space about ourselves . . . may be an idea that could potentially end our existence."*

If They're Out There, They Already Know About Us

Sean Webby

In the following viewpoint, Sean Webby describes what must have been a hugely fun conference held at the SETI (Search for Extraterrestrial Intelligence) Institute in 2010. At that time, efforts to scan for and possibly send signals to potential other life forms were increasing—and not everyone thought that was a good idea. The conference brought together voices on both sides of the issue. At the time he wrote this story, Sean Webby was a reporter for the Mercury News (San Jose).

As you read, consider the following questions:

1. What does Columbus have to do with SETI?
2. What is the main argument for not having projects like SETI?
3. What are some of SETI's goals beyond finding life elsewhere?

"Is Sending Signals to Aliens Really a Good Idea?" by Sean Webby, *Mercury News*, August 15, 2010. Reprinted by permission.

I f we are not alone in the universe, then—considering what happened to the Indians after Columbus landed in America—shouldn't we be keeping a pretty low profile?

That's one of the burning questions in the current scientific search for intelligent extraterrestrial life. The question was debated in front of an enthusiastic crowd of scientists, teachers and amateurs at a SETI Institute conference in Santa Clara over the weekend.

SETI, the Search for Extraterrestrial Intelligence, is a Mountain View-based organization dedicated to leading and enhancing the growing effort to scour the stars for signs that something else living is out there. SETI hopes the conference will become an annual event.

Most agree that the search for intelligent life is a smart idea, which may—if it pays off—revolutionize our existence. But some believe that sending signals and information into space about ourselves, without thoughtfully considering who might be listening, may be an idea that could potentially end our existence.

Award-winning science fiction author Robert Sawyer and SETI scientist John Billingham told the crowd that they agreed with famed scientist Stephen Hawking that there was appropriate cause for concern that some advanced life form could exterminate us. They preached the need for some type of international consultation to at least ponder the risk "of shouting in the jungle."

Seth Shostak, senior astronomer at SETI and the host of the *Are We Alone* radio show, scoffed at the idea, calling it paranoia.

He touched upon the concept that aliens may one day pick up our television signals.

"There is this idea that they see *I Love Lucy* and, because they don't like Fred Mertz's jokes, that therefore they will then destroy the Earth," Shostak said sarcastically. "Now the data we have on alien sociology is very sparse. We have no idea what the aliens might be interested in doing. And if they were going to do something to this planet, this planet has been around for four-and-a-half billion years, why not wipe out the dinosaurs?"

"They did!" someone shouted out from the crowd, to much laughter.

At the conference, SETI sold T-shirts displaying the Drake equation ($N = R^* \, fp \cdot ne \cdot fl \cdot fi \cdot fc \cdot L$)—a formula scientists use to think about the factors that determine how many extraterrestrial civilizations exist in the Milky Way) as well as posters of Spock mind-melding with President Richard Nixon.

The conference openly displayed the often dual nature of SETI's efforts.

SETI is a highly technical and technological scientific effort involving satellites, high-level computer analysis and some of the finest astronomical minds in the world.

On the other hand, most regular folks are far more interested in *Battlestar Galactica* and *Star Trek*. SETI folks understand this and say the conference is part of their effort to teach science, raise money and generate public interest in their quest.

"We know a lot about life here on Earth," SETI Chairman John Gertz said. "We know next to nothing about life elsewhere. That's our mission."

The three-day conference at the Santa Clara Convention Center mixed both popular and technical worlds to address that gap.

There were panel discussions on such topics as dark matter and the Fermi paradox (if there are supposedly so many civilizations in the Milky Way, why haven't we found any hard evidence or been contacted by them?). There was a 1.5-billion-year-old moon rock on display. And there was a talk entitled "Is Doomsday 2012 for real, or will you still have to pay taxes in 2013?"—in addition to a panel where *Star Trek* actors gossiped about their craft.

Star Trek science consultant Andre Bormanis explained how he vetoed an idea to have Capt. Jean-Luc Picard use a high-tech kayak to travel to the molten core of a planet (it takes too long, magma is dangerous, etc.). And he also explained how he lost an argument that no alien society would come to our planet to rob us of our water (the travel costs versus the benefits are way too high).

The thousand or so attendees of the conference were a mix of professional scientists, teachers and students, who saw it as a kind of Lollapalooza of science. College students Erin Laidy and Janine Myszka said they were entranced with the discussions of "multiverses" existing side by side. West Marin high school freshmen Lena McDonnell and Amy Goetz wandered the conference on the sunny weekend with excitement about looking for life in extreme places on earth.

All were agog at being able to brush shoulders with the high-profile scientists.

Nearby, tapping on her laptop in the convention cafe, was Jill Tarter, the award-winning astronomer and SETI research director whose groundbreaking work was fictionalized by Jodie Foster in the movie *Contact*.

Tarter talked about the Allen Telescope Array—partially funded by Microsoft founder Paul Allen—whose intent is to massively expand the ability of scientists to search for signs of life in the universe.

"Will it do the job? I can't answer that question," Tarter said.

"If in fact what I'm supposed to be looking for are Zeta rays and we haven't invented Zeta rays, the best we can do is use the tools and technology we have until we get old enough and wise enough to invent something new."

By the way, her take on the Hawking caution:

"For anybody from 75 to 100 light years away, they have heard the old radio broadcasts. For an older advanced civilization "… they know we are here."

> *"Intentionally signaling other civilizations in the Milky Way Galaxy raises concerns from all the people of Earth, about both the message and the consequences of contact."*

If Aliens Are Watching Us, They May Not Like What They See

Michael P. Oman-Reagan

In the following viewpoint, written six years after the conference described in the previous article, Michael P. Oman-Reagan takes a more serious look at the question of the danger that is potentially posed by reaching out before we know to whom we're reaching out. The author, an anthropologist, points out that when we carefully craft an image of ourselves to broadcast to aliens, we might not be showing our true selves, but what we would like them to see instead. Meanwhile, our broadcasts present a much truer and less flattering image. Michael P. Oman-Reagan is an anthropologist whose scholarly work has focused on human relations with outer space and the search for extraterrestrial intelligence.

As you read, consider the following questions:

1. Seth Shostak, of the SETI Institute, is quoted here as well as in the previous viewpoint. Do you see any similarities in his comments, beyond his position that we should take the risk of reaching out?

2. How does Oman-Reagan apply his experience in anthropology to the question of contacting aliens?

3. If, as the author suggests, an alien civilization might see everything about Earthlings, what do you think they would make of us?

O n March 23, 2016, Microsoft brought a new artificial intelligence (AI) online. Named Tay, she was linked to social media accounts on Twitter, Facebook, and other sites so that she could interact with users and learn how to have conversations with humans. However, Tay soon started to agree with racist and white supremacist posts from social media users and went so far as to deny the Nazi Holocaust, tweet "feminism is cancer," and make transphobic statements. The image of ourselves she reflected was so disturbing that Microsoft shut her down after less than a day online. As *The Telegraph* headline put it, "Microsoft deletes 'teen girl' AI after it became a Hitler-loving sex robot within 24 hours."

Setting aside the nuances of Tay's programming and other factors in the experiment, this story got me thinking about the search for extraterrestrial intelligence (SETI). How might we communicate with alien life from other worlds? What would we even say if we could?

In an episode of the online radio program *StarTalk*, host and astrophysicist Neil deGrasse Tyson spoke with Elon Musk, CEO of the private space corporation SpaceX, about the possible threat of a "superintelligent" AI from space. "Aliens," deGrasse Tyson later wrote, "will make pets of us." Reactions across social media and in the news ranged from calling the comment a lighthearted joke to taking it as a serious assessment of threats waiting for us in interstellar space. The conversation came at a moment when the SETI Institute, a scientific nonprofit dedicated to understanding the origin and nature of life in the universe, had announced renewed interest in "active SETI"—the practice of transmitting messages in the hope of contacting intelligent extraterrestrial life.

Since the Arecibo broadcast of 1974, over 20 transmissions have been attempted by researchers, the public, and private interests, with the aim of contacting extraterrestrial intelligence. In February 2015 at the annual meeting of the world's largest scientific association, the American Association for the Advancement of Science (AAAS), scientists from SETI met to discuss transmitting messages into space. Musk and 27 others responded with a statement on the possible dangers. They used the acronym METI, which stands for "messaging to extraterrestrial intelligence." They warned of "unknown and potentially enormous implications and consequences." And they closed by stating: "Intentionally signaling other civilizations in the Milky Way Galaxy raises concerns from all the people of Earth, about both the message and the consequences of contact. A worldwide scientific, political, and humanitarian discussion must occur before any message is sent." In response, Seth Shostak, senior astronomer and director of the center for SETI research at the SETI Institute, wrote an op-ed in the *New York Times* arguing in favor of active SETI. "The universe beckons," Shostak said, "and we can do better than to declare that future generations should endlessly tremble at the sight of the stars."

Perhaps the Microsoft programmers who created Tay assumed that by interacting with humans online and conducting related internet searches, the artificial intelligence would reflect an idealized image of humanity and how we communicate. Instead Tay showed us the worst of humanity's prejudice, hatred, and bigotry. Tay is a reminder that we may not have as much control over the messages we do send to extraterrestrial life and that we may, in fact, already be sending a message.

Like Tay interacting with people on Twitter, extraterrestrials won't only see what we intend them to see. They will see all of us—

everything we do and say, our entire planet, the full range of what it is to be human. They will see our bigotry, hatred, cooperation, and care—our wars, love, power struggles, artworks, stories, songs, and bombs. Just as Tay didn't see only the good in online discussions, we may not be able to send extraterrestrials a message that represents only the best of our traditions, behaviors, actions, and ideas.

As an anthropologist I've learned there is a significant difference between what people say they do and what people actually do. It sounds like common sense, but it has important implications for anthropology, communication, and SETI. This is one reason that in anthropology we use a research method called "participant observation"—which is really just a fancy way of saying "hanging out with people and paying attention to what they do." If I'm going to meet a group of people and I want to understand what they are up to, what their actions mean, and how they see the world, I could simply ask them what they do. But if I want to really understand what they're actually doing, I have to spend time with them, go along with them in their everyday activities, and listen, watch, and see for myself what they say to each other and how they react when things go differently than expected. An anthropologist has to learn not only what people say they do but also what they really do.

If an extraterrestrial intelligence can understand anything in the specific messages we intentionally send to them, they will probably also be able to understand (or at least perceive) anything else we are doing. Imagine for a moment that we decide to send out a message announcing that we are a peaceful planet with kind intentions and we are interested in cooperation. Looking at Earth, extraterrestrials might see nuclear weapons, fascist governments, surveillance states, racism, intolerance, torture, poverty, hunger, and inequality amid abundance. In other words, they would probably see through our carefully crafted diplomatic lie.

Life elsewhere in the universe may also be profoundly different from life here on Earth. Extraterrestrial intelligence may be almost unrecognizable, with practices, bodies, and forms nothing like our own. After thousands of years of cohabitation and scientific work here on Earth, we have yet to communicate with other intelligent species on our own planet the same way we do with one another. This suggests the challenge of communicating with extraterrestrial life may be at least as great as communicating with other species here on Earth and probably much greater. On the other hand, we may be able to communicate and be understood by simply mutually witnessing one another—by seeing all of their actions and allowing them to see us in return.

By looking out into space one day with telescopes or via spacecraft infinitely more powerful than those we have now, we may be able to watch extraterrestrial beings go about their lives on another world. While we watch them, they may also watch us here on Earth and perhaps also on Mars and other planets in our solar system where we will be living. And by looking at us, they will receive and understand the only message we can send that can be understood and believed: the sum total of all our actions.

In other words, each word I am typing now, every smile you give your neighbor, every television transmission, each walk in the woods, each weapon built, all our internet searches, every injustice committed, every telescope pointed at the sky—the sum total of every human looking up at the stars. This sum total of all our actions may be the message we are already sending to extraterrestrials, a message composed by and reflecting the best and worst of every one of us here on Earth.

> *"To fret about the danger of transmissions to the sky is both too late and too little. Worse, it will endlessly hamstring our descendants."*

Trying to Hide from Aliens Would Do More Harm Than Good

Seth Shostak

In the following viewpoint, Seth Shostak (a scientist who has been quoted in previous viewpoints) argues that worrying about alerting the aliens to our presence through broadcasts is futile. It's already too late. In addition, the author contends, there is no point in limiting our future transmission leakage because such restrictions would drastically prohibit life as we know it. Limiting future leakage also would discourage beneficial technological development for our future. Seth Shostak is an astronomer and director of the search for extraterrestrial intelligence at the SETI Institute in California.

As you read, consider the following questions:

1. To what does the author refer when he writes that the cure is more deadly than the disease?
2. What is our nearest stellar neighbor and how far away is it?
3. What does Einstein's theory of general relativity predict, and why is it useful to the author's argument?

The recent reset of the long-count Maya calendar didn't end the world. But there are serious scientists who worry that Armageddon could soon be headed our way, although from a different quarter—an attack by malevolent, extraterrestrial beings.

The concern is that future radio broadcasts to the stars, intended to put us in touch with putative aliens, might carelessly betray our presence to a warlike society, and jeopardize the safety of Earth. The well-known physicist Stephen Hawking has weighed in on this dreadful possibility, suggesting that we should be careful about sending signals that could trigger an aggressive reaction from some highly advanced race of extraterrestrials.

It all sounds like shabby science fiction, but even if the probability of disaster is low, the stakes are high. Consequently, some cautious researchers argue that it's best to play safe and keep our broadcasts to ourselves. Indeed, they urge a world-wide policy of restraint and relative quiet. They would forbid the targeting of other star systems with transmissions of greater intensity than the routine radio and television that inevitably leak off our planet.

That sounds like a harmless precaution, and who would quibble about inexpensive insurance against the possible obliteration of our world. But this is one worry I don't share. Even more, I believe the cure is more deadly than the disease. To fret about the danger of transmissions to the sky is both too late and too little. Worse, it will endlessly hamstring our descendants.

Ever since the Second World War, we've been broadcasting high-frequency signals that can easily penetrate Earth's ionosphere and seep into space. Many are television, FM radio, and radar. And despite the fact that the most intense of these transmissions sport power levels of hundreds of thousands of watts or more, they dwindle to feeble static at distances measured in light-years. Detecting them requires a very sensitive receiving setup.

As an example of the difficulty, consider an alien society that wields an antenna comparable to the Arecibo telescope in Puerto Rico—a thousand feet in diameter, and the largest single-element radio telescope on Earth. This antenna would be unable to pick

In New Film, Stephen Hawking Warns Against Seeking Out Aliens

"We come in peace" might be the traditional opening gambit for aliens in science fiction, but we should be wary about beaming back a response to any advanced life-forms in real life, Stephen Hawking has warned.

Our first contact from an advanced civilisation could be equivalent to when Native Americans first encountered Christopher Columbus and things "didn't turn out so well," he cautioned.

The comments are made in an online film, *Stephen Hawking's Favorite Places*, in which the theoretical physicist takes viewers on his own CGI spacecraft (the SS *Hawking*) to five significant locations across the cosmos.

On arriving at Gliese 832c, a planet 16 light years away, Hawking reflects: "As I grow older I am more convinced than ever that we are not alone. After a lifetime of wondering, I am helping to lead a new global effort to find out. The Breakthrough Listen project will scan the nearest million stars for signs of life, but I know just the place to start looking. One day we might receive a signal from a planet like Gliese 832c, but we should be wary of answering back."

up our television broadcasts even from the distance of Alpha Centauri which, at 4.4 light-years, is our nearest stellar neighbor. And frankly, it's improbable that we have cosmic confreres this close, or even 10 or 20 times farther. Astronomers such as Frank Drake and the late Carl Sagan have estimated that the nearest Klingons (or whatever their species) are at least a few hundred light-years away. Our leakage signals—when they eventually reach that far—will be orders of magnitude weaker than any our best antennas could detect.

Such arguments might appear to justify the suggestion by the self-appointed defenders of Earth that we need not fear our current broadcasts. They will be undetectably weak. But they claim that we should concern ourselves with deliberate, highly targeted (and

It is not the first time Hawking has warned about the prospect of hostile aliens. Launching the Breakthrough Listen project, which will scan the nearest million stars for signs of life, last year he suggested that any civilisation reading our messages could be billions of years ahead of humans. "If so they will be vastly more powerful and may not see us as any more valuable than we see bacteria," he said.

The 25-minute film, which appears on the platform CuriosityStream, starts at the Big Bang, which has been the focus of much of Hawking's career. Viewers are also taken deep into a super-massive black hole, Sagittarius A*, where Hawking explains his theory of matter, and to Saturn, which Hawking calls "the most spectacular destination in the Solar System."

Finally, Hawking returns to Earth to Santa Barbara where he talks nostalgically of his early career at Cal Tech and times spent on the sunny California coast with his young family.

"My goal is simple: complete understanding of the universe," Hawking said. "It's always been a dream of mine to explore the universe."

"Stephen Hawking Warns Against Seeking Out Aliens in New Film," by Hannah Devlin, Guardian News and Media Limited, September 23, 2016.

therefore highly intense) transmissions. We can continue to enjoy our sitcoms and shopping channels, but we should forbid anyone from shouting in the galactic jungle.

There's a serious flaw in this apparently plausible reasoning. Any society able to do us harm from the depths of space is not at our technological level. We can confidently assume that a culture able to project force to someone else's star system is at least several centuries in advance of us. This statement is independent of whether you believe that such sophisticated beings would be interested in wreaking havoc and destruction. We speak only of capability, not motivation.

Therefore, it's clearly reasonable to expect that any such advanced beings, fitted out for interstellar warfare, will have antenna systems far larger than our own. In the second half of the twentieth century, the biggest of the antennas constructed by earthly radio astronomers increased by a factor of ten thousand in collecting area. It hardly strains credulity to assume that Klingons hundreds or thousands of years further down the technological road will possess equipment fully adequate to pick up our leakage. Consequently, the signals that we send willy-nilly into the cosmos—most especially our strongest radars—are hardly guaranteed to be "safe."

There's more. Von Eshleman, a Stanford University engineer, pointed out decades ago that by using a star as a gravitational lens you can achieve the ultimate in telescope technology. This idea is a straightforward application of Einstein's theory of General Relativity, which predicts that mass will bend space and affect the path of light beams. The prediction is both true and useful: Gravitational lensing has become a favored technique for astronomers who study extremely distant galaxies and dark matter.

However, there's an aspect of this lensing effect that's relevant to interstellar communication: Imagine putting a telescope, radio or optical, onto a rocket and sending it to the Sun's gravitational focus—roughly twenty times the distance of Pluto. When aimed back at the Sun, the telescope's sensitivity will be increased by thousands or millions of times, depending on wavelength. Such an instrument would be capable of detecting even low-power signals (far weaker than your local top-forty FM station) from a thousand light-years. At the wavelengths of visible light, this setup would be able to find the street lighting of New York or Tokyo from a similar remove.

Consequently, it's indisputable that any extraterrestrials with the hardware necessary to engage in interstellar warfare will have the capability to heft telescopes to the comparatively piddling distance of their home star's gravitational focus.

The conclusion is simple: It's too late to worry about alerting the aliens to our presence. That information is already en route at the speed of light, and alien societies only slightly more accomplished than our own will easily notice it. By the twenty-third century, these alerts to our existence will have washed across a million star systems. There's no point in fretting about telling the aliens we're here. The deed's been done, and the letter's in the mail.

But what about a policy to limit our future leakage? What about simply calming the cacophony so we don't continue to blatantly advertise our presence? Maybe our transmissions of the past half-century will somehow sneak by the aliens.

Forget it. Silencing ourselves is both impossible and inadvisable. The prodigious capability of a gravitational lens telescope means that even the sort of low-power transmissions that are ubiquitous in our modern society could be detectable. And would you really want to turn off the radar sets down at the airport, or switch off city streetlights? Forever?

In addition, our near-term future will surely include many technological developments that will unavoidably be visible to other societies. Consider powersats—large arrays of solar cells in orbit around the Earth that could provide us with nearly unlimited energy, sans the noxious emissions or environmental damage. Even in the best cases, such devices would back-scatter hundreds or thousands of watts of radio noise into space. Do we want to forbid such beneficial technologies until the end of time?

Yes, some people are worried about being noticed by other galactic inhabitants—beings that might threaten our lifestyle or even our world. But that's a worry without a practical cure, and the precautions that some urge us to take promise more harm than good. I, for one, have let this worry go.

"Quantum communication is also preferable because of the security it allows for, which is one of the main reasons the technology is being developed here on Earth."

For Safety's Sake, We Should Bring Quantum Science to the Search for ET

Matt Williams

In the following viewpoint, Matt Williams argues that extraterrestrial civilizations may use quantum communication to hide themselves from civilizations they don't want to know they're out there, because using quantum communications would ensure that only extremely advanced civilizations would be able to detect those signals. If we were to develop this technology, we would have a safer way to both search for and communicate with ET. Matt Williams, a freelance writer and science fiction author, is the curator of Universe Today's Guide to Space.

As you read, consider the following questions:

1. What is the origin of quantum communication?
2. How might quantum communication be safer?
3. What are the risks of taking the quantum approach?

"We Could Detect Alien Civilizations Through Their Interstellar Quantum Communication," by Matt Williams, Universe Today, April 28, 2021, https://www .universetoday.com/150900/we-could-detect-alien-civilizations-through-their-interstellar -quantum-communication/. Licensed under CC BY-ND-4.0.

Since the mid-20th century, scientists have been looking for evidence of intelligent life beyond our Solar System. For much of that time, scientists who are engaged in the search for extraterrestrial intelligence (SETI) have relied on radio astronomy surveys to search for signs of technological activity (aka "technosignatures"). With 4,375 exoplanets confirmed (and counting!) even greater efforts are expected to happen in the near future.

In anticipation of these efforts, researchers have been considering other possible technosignatures that we should be on the lookout for. According to Michael Hippke, a visiting scholar at the UC Berkeley SETI Research Center, the search should also be expanded to include quantum communication. In an age where quantum computing and related technologies are nearing fruition, it makes sense to look for signs of them elsewhere.

The search for technosignatures, and what constitutes the most promising ones, has been the subject of renewed interest in recent years. This is due in large part to the fact that thousands of exoplanets are available for follow-up studies using the next-generation telescopes that will be operational in the coming years. With these instruments searching for needles in the "cosmic haystack," astrobiologists need to have a clear idea of what to look for.

In September of 2018, NASA hosted a Technosignatures Workshop, which was followed by the release of their Technosignature Report. By August of 2020, NASA and the Blue Marble Institute sponsored another meeting—Technoclimes 2020—to discuss concepts for future searches that would look for technosignatures beyond the usual radio signals. As someone who has dedicated his professional life to SETI, Hippke has many insights to offer.

The Search Thus Far

As he noted in his study, modern SETI efforts began in 1959 when famed SETI pioneer Giuseppe Cocconi and physicist Philip Morrison (both of Cornell University at the time) published their seminal paper, "Searching for Interstellar Communications." In this paper, Cocconi and Morrison recommended searching for signs of intelligent life by looking for narrow-band signals in the radio spectrum.

This was followed two years later by R. N. Schwartz and C. H. Townes of the Institute of Defense Analyses (IDA) in Washington D.C. In their paper, "Interstellar and Interplanetary Communication by Optical Masers," they proposed that optical pulses from microwave lasers could be an indication of extraterrestrial intelligence (ETI) sending messages out into the cosmos.

But as Hippke notes, six decades and more than one hundred dedicated search programs later, surveys that have looked for these particular technosignatures have yielded nothing concrete. This is not to say that the scientists have been looking for the wrong signatures so far, but that it could be useful to consider casting a wider net. As Hippke explained in his paper:

"We are looking (and should keep looking) for narrow-band lighthouse blasts, even though we have found none yet. At the same time, it is possible to expand our search...It is sometimes argued in the hallways of astronomy departments that we 'just have to tune into the right band' and—voilà—will be connected to the galactic communication channel."

A Quantum Revolution

While virtually all attempts to create quantum processors are relatively recent (occurring since the turn of the century), the concept itself dates back to the early 1970s. It was at this time that Stephen Weisner, a professor of physics at Columbia University at the time, proposed that information could be securely coded by taking advantage of the principle of superposition.

This principle states the "spin" of an electron, a fundamental property that can be oriented "up" or "down," is indeterminate—meaning that it can be either one or both simultaneously. So while an up or down spin is similar to the zeroes and ones of binary code, the superposition principle means that quantum computers can perform an exponentially greater number of calculations at any given time.

Beyond the ability to perform more functions, Hippke identifies four possible reasons why an ETI would opt for quantum communications. These include "gate-keeping," quantum supremacy, information security, and information efficiency. "They are preferred over classical communications with regards to security and information efficiency, and they would have escaped detection in all previous searches," he writes.

The use of computers has evolved considerably over the past century, from isolated machines to the worldwide web, and possibly to an interplanetary network in the future. Looking to the future, Hippke argues that is not farfetched to believe that humanity may come to rely on an interstellar quantum network that enables distributed quantum computing and the transmission of qubits over long distances.

Based on the assumption that humanity is not an outlier, but representative of the norm (aka the Copernican Principle) it is logical to assume that an advanced ETI would have created such a network already. Based on humanity's research into quantum communications, Hippke four possible methods. The first is "polarization encoding," which relies on the horizontal and vertical polarization of light to represent data.

The second method involves the "Fock state" of photons, where a signal is encoded by alternating between a discreet number of particles and vacuum (similar to binary code). The two remaining options involve either time-bin encoding—where early and late arrival is used—or coherent state of light encoding, where light is amplitude-squeezed or phase-squeezed to simulate a binary code.

Security and Supremacy

Of the many benefits that quantum communications would present for a technologically advanced species, Gate-Keeping is especially interesting because of the implications it could have for SETI. After all, the disparity between what we assume is the statistical likelihood of intelligent life in our Universe and the lack of evidence for it (aka the Fermi Paradox) cries out for explanations. As Hippke puts it:

"ETI may deliberately choose to make communications invisible for less advanced civilizations. Perhaps most or all advanced civilization feel the need to keep the "monkeys" out of the galactic channel, and let members only participate above a certain technological minimum. Mastering quantum communications may reflect this limit."

The idea of quantum communication was first argued by Mieczyslaw Subotowicz, a professor of astrophysics at the Maria Curie-Sklodowska University in Lublin (Poland), in 1979. In a paper titled "Interstellar communication by neutrino beams," Subotowicz argued that the difficulties this method presented would be a selling point to a sufficiently advanced extraterrestrial civilization (ETC).

By opting for a means of communication that has such a small cross-section, an ETC would only be able to communicate with similarly advanced species. However, Hippke noted, this also makes it virtually impossible to detect entangled pairs of neutrinos. For this reason, entangled photons would not only provide for gate-keeping, but they would also be detectable by those meant to receive them.

Similarly, quantum communication is also preferable because of the security it allows for, which is one of the main reasons the technology is being developed here on Earth. Quantum key distribution (QKD) enables two parties to produce a shared key that can be used to encrypt and decrypt secret messages. In theory, this will lead to a new era where encrypted communications and databases are immune to conventional cyber attacks.

In addition, QKD has the unique advantage of letting the two parties detect a potential third party attempting to intercept their messages. Based on quantum mechanics, any attempt to measure a quantum system will collapse the wave function of any entangled particles. This will produce detectable anomalies in the system, which would immediately send up red flags. Said Hippke:

> We do not know whether ETI values secure interstellar communication, but it is certainly a beneficial tool for expansive civilizations which consist of actions, like humanity today. Therefore, it is plausible that future humans (or ETI) have a desire to implement a secure interstellar network.

Another major advantage to quantum computing is its ability to solve problems exponentially faster than its digital counterparts —what is known as "quantum supremacy." The classic example is Shor's algorithm, a polynomial-time quantum algorithm for factoring integers that a conventional computer would take years to solve, but a quantum computer could crack in mere seconds.

In traditional computing, public-key encryption (such as the RSA-2048 encryption) employs mathematical functions that are very difficult and time-consuming to compute. Given that they can accommodate an exponentially greater number of functions, it is estimated that a quantum computer could crack the same encryption in about ten seconds.

Last, but not least, there's the greater photon information efficiency (PIE) that quantum communications offer over classical channels—measured in bits per photon. According to Hippke, quantum communications will improve the bits per photon efficiency rating by up to one-third. In this regard, the desire for more efficient data transmissions will make the adoption of a quantum network something of an inevitability.

"Turned the other way around, classical channels are energetically wasteful, because they do not use all information encoding options per photon," he writes. "A quantum advantage of order 1/3 does not seem like much, but why waste it? It is logical to assume that ETI prefers to transmit more information rather than less, per unit energy."

Challenges

Of course, no SETI-related pitch would be complete without mentioning the possible challenges. For starters, there's the matter of decoherence, where energy (and hence, information) is lost to the background environment. Where transmissions through interstellar space are concerned, the main issues are distance, free electrons (solar wind), interplanetary dust, and the interstellar medium—low-density clouds of dust and gas.

"As a baseline, the largest distance over which successful optical entanglement experiments have been performed on Earth is 144 km," notes Hippke. Since the mass density of the Earth's atmosphere is 1.2 kg m-3, this means that a signal passing through a column 144 km (~90 mi) in length was dealing with a column density of 1.728×105 kg m-2. In contrast, the column density between Earth and the nearest star (Proxima Centauri) is eight orders of magnitude lower (3×10-8 kg m-2).

Another issue is the delay imposed by a relativistic Universe, which means that messages to even the closest star systems would take years. As a result, quantum computation is something that will be performed locally for the most part, and only condensed qubits will be transmitted between communication nodes. With this in mind, there are a few indications humanity could be on the lookout for in the coming years.

What to Look For?

Depending on the method used to transmit quantum information, certain signatures would result that SETI researchers could identify. At present, SETI facilities that conduct observations in the visible light spectrum are not equipped to receive quantum communications (since the technology does not exist yet). However, they are equipped to detect photons, obtain spectra, and perform polarization experiments.

As such, argues Hippke, they would be able to tease out potential signals from the background noise of space. This is similar to what Professor Lubin suggested in a 2016 paper ("The Search for Directed Intelligence"), where he argued that optical signals (lasers) used for directed-energy propulsion or communications would result in occasional "spillover" that would be detectable.

In much the same way, "errant" photons could be collected by observatories and measured for signs of encoding using various techniques (including the ones identified in the study). One possible method Hippke recommends is long-duration interferometry, where multiple instruments monitor the amplitude and phase of electromagnetic fields in space over time and compare them to a baseline to discern the presence of encoding.

One thing bears consideration though: If by listening in on ETI quantum communications, won't that cause information to be lost? And if so, would the ETI in question not realize we were listening in? Assuming they were not aware of us before, they sure would be after all this went down! One might conclude that it would be better to not eavesdrop on the conversations of more advanced species!

But that's a question for another day and another fertile topic for debate.

> *"If they wanted to invade and destroy our planet, they would have done it millions of years ago when early life made it obvious this planet was inhabited."*

What Will We Do If We Actually Make Contact?

Fraser Cain

In the following somewhat lighthearted viewpoint, Fraser Cain sums up many of the issues that have been discussed previously in this chapter and in this book, but with a twist. Cain asks: What if we do make contact with another intelligent civilization? What then? However we choose to proceed in the event of contact, Cain doesn't seem too worried. The author believes that if alien beings knew about us, they would have tried to make contact already. Fraser Cain is the publisher of Universe Today.

As you read, consider the following questions:

1. What is the difference between SETI and CETI?
2. What is the Rio Scale?
3. What does Seth Shostak suggest sending the aliens to help them understand what Earthlings are all about?

"What If We Do Find Aliens?" by Fraser Cain, Universe Today, November 24, 2016, https://www.universetoday.com/132102/what-if-we-do-find-aliens/. Licensed under CC BY 4.0 International.

Time to talk about my favorite topic: aliens.

We've covered the Fermi Paradox many times over several articles on Universe Today. This is the idea that the Universe is huge, and old, and the ingredients of life are everywhere. Life could and should have appeared many times across the galaxy, but it's really strange that we haven't found any evidence for them yet.

We've also talked about how we as a species have gone looking for aliens. How we're searching the sky for signals from their alien communications. How the next generation of space and ground-based telescopes will let us directly image the atmospheres of extrasolar planets. If we see large quantities of oxygen, or other chemicals that shouldn't be around, it's a good indication there's life on their planet.

We've even talked about how aliens could use that technique on us. We've been sending our radio and television signals out into space for the last few decades. Who knows what crazy things they think about our "historical documents"? But Earth life itself has been broadcasting our existence for hundreds of millions of years, since the first plankton started filling our atmosphere with oxygen. A distant civilization could be analyzing our atmosphere and know exactly when we entered the industrial age.

But what we haven't talked about, the space elephant in the room, if you will, is what we'll do if we actually make contact. What are we going to say to each other? And what will happen if the aliens show up?

Although there's no official protocol on talking to aliens, scientists and research institutions have been puzzling out the best way we might communicate for quite a while.

Perhaps the best example is the SETI Institute, the US-based research group who have dedicated radio telescopes scanning the skies for messages from space.

Let's imagine you're a SETI researcher, and you're browsing last night's logs and you see what looks like a message. Maybe it's instructions to build some kind of dimensional portal, or a recipe book.

Whatever you do, don't try out the recipes. Instead, you need to make absolutely sure you're not dealing with some kind of natural phenomenon. Then you need to reach out to other researchers and get them to confirm the signal.

If they agree it's aliens, then you need to inform the International Astronomical Union and other international groups, like the United Nations, Committee on Space Research, etc.

Unless they've got some good reason to stop you, it's time to announce the discovery to the worldwide media. You made the discovery, you get to break the news to the world.

At this point, of course, the entire world is going to freak right out. Whatever you do, however, you have to resist the urge to send back a message or build that dimensional portal, no matter how much you think you understand the science. Instead, let an international committee mull it over while you stockpile supplies in a secret alien proof bunker in the desert.

What kind of message should we actually craft to our new alien penpals? Will we become fast friends, jump starting our own technological progress, or will we insult them by accident?

In 2000, an international group of SETI researchers including the famous Jill Tarter devised the Rio Scale. It's really easy to use, and there's even a fun online calculator.

Step 1, figure out the class of phenomenon. Is it a message sent directly to Earth, expecting a reply? Or did we merely find some alien artifact or old timey Dyson sphere orbiting a nearby star?

Step 2, how verifiable is the discovery? Are we talking ongoing signals received by SETI researchers, or a hint in some old data that's impossible to confirm?

Step 3, how far are we talking here? Hovering over Paris? Within our Solar System, or outside the galaxy?

Step 4, how sure are you? 100% certain, and everyone agrees because they can all see that enormous mothership floating above London? Or nobody believes you, and they've locked you up because of your insane ramblings and misappropriation of government equipment?

Punch in your numbers and you'll get a rank on the Rio Scale between 0 and 10. Level 0 is "no importance" or "you're a crank," while level 10 is "extraordinary importance," or "now would be a good time to panic."

SETI researcher Seth Shostak, calculated the Rio Scale for various sci-fi movies and shows. The first message from aliens in *Independence Day* would count as a 4. While the obliteration of the White House by a massive floating alien city that everybody could see would count as a 10.

The messages received in *Contact*, and independently confirmed by researchers around the world, would qualify in the 4-8 range, while the monolith discovered on the Moon in 2001 would be a solid 6.

Now you know how important the discovery is, what do you say back to those chatty aliens?

This falls under the term CETI, which means Communications with Extraterrestrial Aliens, which shouldn't be confused with SETI, or the Search for Extraterrestrial Aliens. And it turns out, that horse has already left the stable.

When the Pioneer and Voyager spacecraft were constructed, they were equipped with handy maps to find Earth's precise location in the Milky Way.

In 1974, Carl Sagan and Frank Drake composed a message in alienese and broadcast it into space from the Arecibo Observatory.

In 1999 and 2003 a series of signals were transmitted towards various interesting stars. The messages contained images of Earth, as well as various mathematical principles that could be used by aliens as a common language.

We'll know if that was a good idea in a few decades.

In 2015, scientists like David Grinspoon, Seth Shostak and David Brin collected together to discuss if it's a wise idea to send messages off into space, to broadcast our existence to potentially hostile alien civilizations.

According to Seth Shostak, the best message we can send is the entire internet. Just send it all, they'll work out what we're all about.

The science fiction author David Brin thinks that's a terrible idea, and we should keep our mouths shut.

Personally, I think the aliens already know we're here. If they wanted to invade and destroy our planet, they would have done it millions of years ago when early life made it obvious this planet was inhabited. The jig is up.

It's a mind bending concept to imagine what life might be like if we knew with absolutely certainty that there's an alien civilization right over there, on that world. I'm sure people will freak out for a while, but then we'll probably just go back to life as normal. Human beings can get bored by the most surprising and amazing things.

If you learned there was definitely an alien civilization out there, how do you think humanity would respond?

Periodical and Internet Sources Bibliography

The following articles have been selected to supplement the diverse views presented in this chapter.

Rebecca Boyle, "Why These Scientists Fear Contact with Space Aliens," NBC News, February 8, 2017. https://www.nbcnews .com/storyline/the-big-questions/why-these-scientists-fear -contact-space-aliens-n717271

Liz Braun, "Maybe Communicating with Extraterrestrials Is a Bad Idea," *Toronto Sun*, June 14, 2021. https://torontosun.com/news /maybe-communicating-with-extraterrestrials-is-a-bad-idea

Mark Buchanan, "Contacting Aliens Could End All Life on Earth. Let's Stop Trying," *Washington Post*, June 13, 2021. https://www .washingtonpost.com/outlook/ufo-report-aliens-seti/2021/06/09 /1402f6a8-c899-11eb-81b1-34796c7393af_story.html

Fraser Cain, "Are Aliens Watching Old TV Shows?" Universe Today, January 19, 2015. https://www.universetoday.com/118250/are -aliens-watching-old-tv-shows/

Steven Johnson, "Greetings ET (Please Don't Murder Us)," *New York Times*, June 28, 2017. https://www.nytimes.com/2017/06/28 /magazine/greetings-et-please-dont-murder-us.html

Arik Kershenbaum, "What Do Aliens Look Like? Animals on Earth May Hold the Answer," Science Focus, November 26, 2020. https://www.sciencefocus.com/science/what-do-aliens-look-like/

Ilima Loomis, "Should We Call Out to Space Aliens?" Science News for Students, March 21, 2017. https://www .sciencenewsforstudents.org/article/should-we-call-out-space -aliens

Adam Mann, "Want to Talk to Aliens? Try Changing the Technological Channel Beyond Radio," *Scientific American*, September 2, 2020. https://www.scientificamerican.com/article /want-to-talk-to-aliens-try-changing-the-technological-channel -beyond-radio/

Dennis Overbye, "Was That a Dropped Call from ET?" *New York Times*, December 31, 2020. https://www.nytimes .com/2020/12/31/science/radio-signal-extraterrestrial.html

Passant Rabie, "To Find Aliens, Scientists Are Hunting for Interstellar Encrypted Messages," Inverse, August 7, 2021. https://www.inverse.com/science/quantum-communication-could-reveal-aliens

Sara Rigby, "Should We Be Signalling Our Existence to Alien Life?" Science Focus, January 14, 2021. https://www.sciencefocus.com/space/should-we-be-signalling-our-existence-to-alien-life/

Ker Than, "Stanford's Scott Hubbard Contributed to New 'Planetary Quarantine' Report Reviewing Risks of Alien Contamination," *Stanford News*, May 7, 2020. https://news.stanford.edu/2020/05/07/new-planetary-quarantine-report-reviews-risks-alien-contamination-earth/

Jacco van Loon, "Aliens: Could Light and Noise from Earth Attract Attention from Outer Space?" The Conversation, July 31, 2019. https://theconversation.com/aliens-could-light-and-noise-from-earth-attract-attention-from-outer-space-121073

OPPOSING
VIEWPOINTS®
SERIES

CHAPTER 4

Is the Search for Life Elsewhere in the Universe a Waste of Resources?

Chapter Preface

So far we've read viewpoints debating whether or not the search for aliens is worth the trouble (Are there any out there?), whether it is too late (Are they already here?), and whether it is prudent (Could they kill us?). Despite all those questions, one important concern remains: cost. Searching for extraterrestrial intelligence is not cheap. In this chapter, the authors wonder if searching for life beyond Earth is worth the money.

In a world that has so many problems, costly programs to find intelligent life on other planets might seem like an indulgent waste of resources. Surely that money could be put to one or many better uses. On the other hand, some experts make the case that the benefits we get from a robust space exploration, including the search for ET, are well worth the cost and could actually contribute to solving some of the problems we face here, such as poverty and war.

This chapter opens with a discussion of how SETI (search for extraterrestrial intelligence) programs have been funded and defunded over the years and how the programs have always struggled for money, sometimes even relying on small-dollar donations from ordinary citizens. A billionaire scientist has stepped in to help, which may or may not be a good thing. Later viewpoints dig deep into the primary question of this chapter: Is spending money exploring space and looking for aliens not only a waste of time and money but also a waste that costs humans and our societies dearly by diverting funds from much worthier and more urgent projects?

As we've seen throughout this volume, there is plenty of debate about many aspects of the search for alien life. Here the authors ask: Should we be doing this at all?

| "*[Milner] knew of SETI's dire financial straits, and he believed his money and knowledge of the tech industry could help speed up the search.*"

Private Money Makes the Search for ET a Better Bargain

Daniel Clery

In the following viewpoint, Daniel Clery delves into SETI's $100 million Breakthrough Listen initiative. The author explains where the money for that came from and how that surge in funding has completely revitalized the search for extraterrestrial intelligence and brought it into the mainstream. That means new, huge possibilities, but a few risks as well. Daniel Clery is a British science writer who specializes in physics, astronomy, space science, and European space policy.

As you read, consider the following questions:

1. How did the financial crisis of 2008 affect SETI?
2. Why did Yuri Milner choose to invest in science, and SETI in particular?
3. How, according to sources cited here, might this windfall of funding distort the science?

"How Big Money Is Powering a Massive Hunt for Alien Intelligence," by Daniel Clery, *Science*, September 10, 2020. Reprinted by permission.

In 2015, Sofia Sheikh was at loose ends. Her adviser at the University of California (UC), Berkeley, with whom she studied hot, giant exoplanets, had left for a new job. Browsing reddit, she saw a post about a lavishly funded new search for extraterrestrial intelligence (SETI) and noticed that its leader was also at UC Berkeley: astrophysicist Andrew Siemion. She asked her former adviser for an introduction and met with Siemion when he was still unpacking boxes in a new office. "Everything's kind of history from there," says Sheikh, who became the team's first undergraduate student.

Sheikh is now a Ph.D. student at Pennsylvania State University (Penn State), University Park, where she led a radio survey of 20 nearby star systems aligned with Earth's orbital plane. If an intelligent civilization inhabited one of these systems and pointed a powerful telescope our way, they would see Earth passing in front of the Sun, and they might detect signs of life in our atmosphere. They might even decide to send us a message. The results, published in February in *The Astrophysical Journal*, were unsurprising. "Spoiler alert: no aliens," Sheikh jokes.

SETI researchers are used to negative results, but they are trying harder than ever to turn that record around. Breakthrough Listen, the $100 million, 10-year, privately funded SETI effort Siemion leads, is lifting a field that has for decades relied on sporadic philanthropic handouts. Prior to Breakthrough Listen, SETI was "creeping along" with a few dozen hours of telescope time a year, Siemion says; now it gets thousands. It's like "sitting in a Formula 1 racing car," he says. The new funds have also been "a huge catalyst" for training scientists in SETI, says Jason Wright, director of the Penn State Extraterrestrial Intelligence Center, which opened this year. "They really are nurturing a community."

Breakthrough Listen is bolstering radio surveys, which are the mainstay of SETI. But the money is also spurring other searches, in case aliens opt for other kinds of messages—laser flashes, for example—or none at all, revealing themselves only through passive "technosignatures." And because the data gathered by Breakthrough Listen are posted in a public archive, astronomers are combing through it for nonliving phenomena: mysterious deep-space pulses called fast radio bursts and proposed dark matter particles called axions. "There are untapped possibilities here," says axion searcher Matthew Lawson of Stockholm University.

Perhaps the most important consequence of Breakthrough Listen is that it has nudged SETI, once considered fringe science, toward the mainstream. "Journals are relaxing and letting good technosignature papers be published," says astrobiologist Jacob Haqq-Misra of the Blue Marble Space Institute of Science. "The giggle factor is reducing." After nearly 3 decades of eschewing SETI, NASA organized a technosignature workshop in 2018. In June, it awarded a grant to model the detectability of possible technosignatures in the atmospheres of exoplanets, its first ever SETI-related grant not involving radio searches.

But some astronomers worry the funding boon is distorting science. Fernando Camilo, chief scientist of the South African Radio Astronomy Observatory, says Breakthrough Listen's voracious appetite for time on large telescopes leaves him uncomfortable. "It leaves less time to do astronomy." Others say SETI's high-risk, rush-for-the-prize approach could distract funders from a more rational, stepwise search for extraterrestrial life. "We do have a really thoughtful process on what gets funded and what doesn't," says Harvard University astronomer David Charbonneau. "That doesn't happen with rich individuals."

But SETI proponents don't see themselves as separatists. They are increasingly working hand in hand with those searching for exoplanets and studying astrobiology. "Looking for intelligence is the logical conclusion of this search for life," says astronomer David Kipping of Columbia University.

SETI started small. In 1960, astronomer Frank Drake pointed a 26-meter radio telescope in Green Bank, West Virginia, at two nearby Sun-like stars. He scanned frequencies around

1.42 gigahertz, which correspond to wavelengths of about 21 centimeters—the part of the spectrum where clouds of interstellar hydrogen emit photons. This 21-centimeter glow is ubiquitous, and Drake supposed it might be a universal channel on the cosmic dashboard, a natural place for a clarion "We are here!" But his targets, Tau Ceti and Epsilon Eridani, were expressionless. The survey, called Project Ozma, saw no sign of artifice, such as an intense spike squeezed into a narrow frequency band.

With funding from NASA and the National Science Foundation (NSF), however, searches continued, with bigger telescopes to listen for fainter signals and hardware that could scan thousands and eventually millions of narrow frequency channels at once. Drake devised his now famous, eponymous equation that estimates how many communicative extraterrestrial civilizations may exist in the Milky Way. It depends on seven variables, from the rate of star formation to the average lifetime of a civilization. Even though only one of the seven factors—star-formation rate—was known with any certainty, alien hunters were on the prowl.

In 1992, NASA decided to look harder, only to quickly reverse course. It embarked on the Microwave Observing Project, a 10-year, $100 million SETI search using several large telescopes. But the following year, the project was ridiculed and cut by lawmakers focused on reducing the federal budget deficit. Ever since, NASA has mostly shied away from SETI.

Even as federal funding shriveled, the 1990s gave SETI an unexpected gift. Until then no one had detected an exoplanet, much less a potentially hospitable one, but that decade brought a host of discoveries. Since then, missions such as NASA's Kepler telescope have suggested that planetless stars are rare, and that about one in five Sun-like stars has potentially habitable Earth-size planets—two more factors in the Drake equation that have fueled optimism among SETI advocates. The turn-of-the-century tech boom offered another boost: newly minted billionaires with a taste for space. A high point came in 2007 with the inauguration of the Allen Telescope Array, a SETI observatory in California kick-started with $11.5 million from Microsoft cofounder Paul Allen.

Then the field took another plunge. The 2008 financial crisis struck and within a few years, with federal and state funding tight, UC Berkeley withdrew from the project. The array was put into hibernation for 8 months. A planned expansion from 42 to 350 dishes never materialized. "SETI was entirely decimated," Siemion says. "I was one of maybe two or three in the whole world working on SETI."

That was when Yuri Milner called.

Born and educated in Moscow, Milner worked as a particle physicist at the Lebedev Physical Institute. In 1990, as the Soviet Union collapsed, he left to study business at the University of Pennsylvania, and in 1999 he founded an internet investment fund. The fund was an early backer of Facebook and Twitter, and later Spotify and Airbnb. *Forbes* magazine puts Milner's net worth at $3.8 billion. "I made some lucky investments," he tells *Science*.

Milner says he's always felt a connection with space and SETI. He was born in 1961, days after Drake convened the first SETI conference. He is named after Yuri Gagarin, the first cosmonaut. Once he had built up a fortune, "I discovered that now I can give back to science," he says. He knew of SETI's dire financial straits, and he believed his money and knowledge of the tech industry could help speed up the search. Siemion's UC Berkeley center, across the San Francisco Bay from Milner's home in Silicon Valley, became the beneficiary.

Breakthrough Listen set out ambitious goals. It would survey 1 million of the closest stars to Earth and 100 nearby galaxies using two of the world's most sensitive steerable telescopes, the 100-meter Green Bank Telescope in West Virginia and the

64-meter Parkes radio telescope in Australia. Buying up about 20% and 25% of the time on those telescopes, Breakthrough Listen promised to cover 10 times more sky than previous surveys and five times more of the radio spectrum, and gather data 100 times faster.

Achieving these goals required new hardware. The key electronic component is a digital backend, which chops telescope data into ultrathin frequency slices and records it. Siemion says Breakthrough Listen's backends are "orders of magnitude more powerful than anything else on site." The instruments are available for 100 hours every year to other astronomers interested in such fine frequency resolution. That allocation is often oversubscribed at Green Bank, Siemion says, ever since the backend helped characterize the first repeating fast radio burst.

The project is adding a major new telescope to its mix of collaborations: MeerKAT, a South African array of 64 dishes each 13.5 meters across. Instead of buying time on the array, Breakthrough Listen is tapping into the data stream while the telescope observes its regular targets—a procedure known as commensal observing. "You take what you can get," Camilo says. "When it works, it's fantastic." Commensal observing will also be added to the Karl G. Jansky Very Large Array in New Mexico, the workhorse of US radio astronomy, in a project led by the privately funded SETI Institute.

Gathering data sets is one thing; scouring heaps of them for alien messages is another. SETI researchers have long looked for energy packed into narrow frequency signals—something that is hard for nature to replicate, although astronomers need to exclude humanmade signals. One test is to see whether the signal's frequency drifts over time: An alien transmitter would be on a moving planet, causing a Doppler shift. If the frequency is rock steady, it's likely to be earthly interference. Similarly, if the signal persists when the telescope moves from its target, it's noise from Earth.

But aliens might send something more complex than a single loud note. How do you scan SETI data for something that just seems anomalous or weird? Researchers have been trying to enlist artificial intelligence (AI), but it hasn't been easy. One species of AI, natural language algorithms, can recognize key words in the flow of human speech—think of Amazon's Alexa, or eavesdroppers at the National Security Agency—after being trained on vast speech data sets. But the huge number of narrow frequency channels in SETI data overwhelms these algorithms.

Converting the data stream into 2D diagrams that resemble images works better, at least in tests, in which machine vision algorithms picked out strange pictures from a torrent of similar ones. "We have to guess what an anomaly might look like and train the algorithm to look for this, or look for things that look similar," says Steve Croft of UC Berkeley's SETI Research Center.

The focus of SETI searches tends to reflect the technology of the times. Radio was in its heyday when Drake started out. But as lasers have grown in power and sophistication, so have efforts to spot alien laser signals with so-called optical SETI.

Astronomers have carried out optical searches with modest telescopes since the 1990s. Breakthrough Listen is doing its own, with time on the 2.4-meter Automated Planet Finder (APF) telescope at the Lick Observatory in California. APF has been scanning a sample of stars to distances up to 160 light-years but will now work through a new list: stars with potentially habitable planets identified by NASA's Transiting Exoplanet Survey Satellite.

Others are developing telescopes that wouldn't need to target individual stars. The LaserSETI project, funded by the SETI Institute, is a collection of $30,000 miniobservatories, made up of an off-the-shelf fisheye lens, two cameras, and electronics that would gather light from the entire sky. The first was installed last year on an observatory roof north of San Francisco. Eventually, the institute wants to install 60 instruments around the world for 24/7 coverage.

LaserSETI's small telescopes would only pick up an especially bright flash from a nearby source. Shelley Wright of UC San Diego hopes to see much farther with the Pulsed All-sky Near-infrared Optical SETI (PANOSETI), an all-sky telescope able to detect ultrashort laser pulses across all optical wavelengths.

PANOSETI's design includes lightning-fast photon counters sensitive to pulses less than one-billionth of 1 second long. "It's hard for nature to make that," Shelley Wright says. It relies on a Fresnel lens, a type used in lighthouses to focus light into a narrow beam. Flipped over, a Fresnel can gather light from a 10°-wide patch of sky onto the photon counters. The team is building two observatories, each an array of 80 telescopes with lenses 50 centimeters across, bunched together in a fly's eye arrangement. The plan is to site the pair 1 kilometer apart—to help root out false positives—at the Palomar Observatory in California. Funded by Qualcomm cofounder Franklin Antonio, the project has built five telescopes but has been stalled by the COVID-19 pandemic.

Then again, even intelligent aliens might be too busy or too shy to send messages to the stars. So SETI researchers also hope to detect passive signs of technology. People's ideas about what to look for often reflect their time: Consider the 19th century "discovery" of canals on Mars when canals were still a common form of transport on Earth. In 1960, amid rapid economic growth and concerns about energy shortages, physicist Freeman Dyson imagined an advanced society might build a megastructure surrounding a star to capture its energy. Such "Dyson spheres" continue to fascinate and were suggested as an explanation for the strange dimmings of the star KIC 8462852, known as Tabby's Star. In 2015, Jason Wright led a search for the glow of Dyson spheres in 100,000 nearby galaxies, using data from NASA's Wide-field Infrared Survey Explorer satellite.

Technosignatures could be more subtle. In the not-too-distant future, ultrasensitive radio telescopes might be able to pick up the beams of a radar, like the ones used for air traffic control, from a distant exoplanet. Future optical telescopes might reveal the glow of a city's lights or its infrared warmth. Heavy industry or geoengineering might leave fingerprints in a planet's atmosphere.

These efforts chime with searches for biosignatures, detectable marks that organic life might leave on an exoplanet. "The line between technosignatures and biosignatures is blurring," Sheikh says. "It makes sense to observe both." In deciding to fund the 2018 workshop on technosignatures, NASA felt that they could be discussed "on a firmer scientific foundation than before," says Michael New, the agency's deputy associate administrator for research. After the workshop, the wording in NASA funding calls that had for some years excluded SETI-related proposals quietly disappeared.

In June, Jason Wright and his colleagues benefited from the new openness when they were awarded a grant to model exoplanet atmospheres and put together a "library" of potential technosignatures, which astronomers can refer to when observing exoplanets. The team will first model chlorofluorocarbons—a pollutant that isn't produced naturally—and vast solar power arrays, because they would leave an obvious cutoff in the ultraviolet part of the spectrum. "What we should look for is things that can't be avoided, civilization's manifestations in the biosphere," says Adam Frank, lead investigator on the grant at the University of Rochester.

But even after the fanfare of Breakthrough Listen, SETI remains far from a central concern for most astronomers. In 2018, panels of researchers convened by the National Academies of Sciences, Engineering, and Medicine (NASEM) drew up strategies for NASA on astrobiology and exoplanets. They made scant mention of technosignatures and didn't advise NASA to spend any money on the topic, or, more generally, SETI.

SETI enthusiasts say they are trying to avoid being shut out of an even bigger NASEM effort: its decadal survey of astrophysics, a once-a-decade priority setting exercise that is influential with funding agencies and legislators. The survey is due to report early next year. "We've made a big push to get the decadal survey ... to explicitly say that NASA and the NSF need to nurture this field," Jason Wright says. He and colleagues made nine submissions, known as white papers, to the survey, compared with a single white paper in the previous survey. Sheikh says: "There are signs the winds are starting to shift."

But many astronomers think the more important hunt is for alien life of a more basic kind, not the higher risk search for technological societies. "We have to invest in general questions," says Charbonneau, who co-chaired the NASEM panel that developed the NASA exoplanet hunting strategy. "If we just go for the prize and don't find anything, what have we learned from that?"

Mainstream astrobiologists hope the decadal survey will give a thumbs up to the Large UV/Optical/IR Surveyor, or LUVOIR, a proposed NASA space telescope as much as six times wider than the Hubble Space Telescope. It would scrutinize habitable planets for biosignatures and estimate the fraction of them that support life—another term in the Drake equation. "The progress we've made as scientists follows the terms of the Drake equation in order," says astrobiologist Shawn Domagal-Goldman of NASA Goddard Space Flight Center. "That progress could lead to a search for technosignatures. I could see LUVOIR being used to do that, even though it wasn't designed for such a search."

Jason Wright, however, thinks the potential payoff of SETI is just too tempting to put off the search. In July, he and his colleagues reported the "discovery space"—all the possible locations, frequencies, sensitivities, bandwidths, timings, polarizations, and modulations—that SETI radio surveys have so far explored. The result: If the entire discovery space is represented by the world's oceans, SETI has so far searched the volume of a hot tub.

Milner seems ready to support at least a few more SETI hot tubs. He says he wants Breakthrough Listen to continue past 2025, when his initial funding runs out. "It's one of the most existential questions in our universe," he says. "Just knowing we are not alone...is something that can bring us together here on Earth."

> *"I think that SETI programmes are probably doomed to fail—although I would love to be proved wrong."*

SETI-Style Searches Are a Waste of Time and Money

Andrew Norton

In the following viewpoint, Andrew Norton opens with what is by now the familiar discussion of the Fermi paradox and the Drake equation and sets up the basic issue that was debated in chapter 1: What are the odds of life on other planets? This author thinks the odds are quite low, meaning that SETI's Breakthrough Initiative project is a waste of $100 million. However, Norton does have another suggestion about how we might go about the search and what we are likely to find. Andrew Norton is professor of astrophysics education at the Open University and former vice president of the Royal Astronomical Society.

As you read, consider the following questions:

1. What are the three possibilities Norton lists for why we have not heard from an alien civilization?
2. Why does he say aliens might be hiding?
3. What method of searching for life might be worthwhile?

The launch of the $100M Breakthrough Initiative project to Search for Extraterrestrial Intelligence (SETI) has been supported by many leading scientists including Stephen Hawking and astronomer royal Martin Rees. But there is no evidence—and few convincing theories—to suggest that intelligent, communicative aliens actually exist. So are listening projects really the best way to search for extraterrestrial life?

The possibility of life outside our own planet has been the subject of debate for centuries, with the essence of the problem crystallised by Italian physicist Enrico Fermi in 1950. His now famous "Fermi paradox" runs simply: if intelligent life exists elsewhere in the Galaxy, then why do we see no evidence for it?

Colonising the Galaxy—Hard but Possible

We now know that planets around other stars are very common. Since the first discovery of a planet orbiting the star 51 Pegasi in 1995, around 2000 exoplanets have now been found. Most of these are close by—within a few hundred light years.

Statistical analysis of the results from the Kepler spacecraft suggest that as many as one-fifth of all sun-like stars has an Earth-like planet in its habitable zone, where conditions are such that liquid water could exist.

So if planets are so plentiful, then what about life? The Drake equation, formulated by Frank Drake in 1961, attempts to answer this question by suggesting there could be many civilisations in the Milky Way that we should be able to communicate with.

However, while many of the terms in the equation are now known fairly well, others are highly uncertain. But let's assume for a moment that such civilisations do exist. If they do, then might we notice them? A straightforward way for an alien civilisation to make itself known is simply to colonise the galaxy. Let's consider how long this might take, assuming technology that is not too far away.

It would be possible now to build probes that could be sent out into space to search for other planets, land on them, and build

THE SEARCH IS WORTH THE MONEY BECAUSE IT BRINGS US TOGETHER

It is estimated that for humans to have first reached Australia approximately 65,000 years ago, they would have had to travel across vast bodies of open water. Much is unknown about how exactly they achieved this. Archaeologists and historians credit this as one of the first, remarkable achievements made by humans. While admirable today, their courage could also be described as reckless.

When these early humans set sail, they took incalculable risks. They had to weather the scorching sunlight that threatened them with dehydration. They had to accept sharp pangs of hunger due to a lack of assured food sources. They had to take on the challenges of navigation with primitive rafts… all with zero assurance that anything would meet them at their journey's end.

Some people ask why we should waste money on space exploration when there are so many problems on Earth. The main answer seems to be embedded in our genetic code.

To explore is to be human.

Even if we have no assurance of finding what we seek, the drive to discover the unknown is what has defined us as a species. Space missions are the next step in our evolution of exploration.

replicas of the probe that could in turn be sent out to other planets and so on.

At the sort of speeds we can now imagine, such as that achieved by the New Horizons spacecraft (60,000 km/h), it would take a mere 18,000 years to travel a distance of one light-year. Let's assume such a probe were sent to a planet ten light-years away, arriving after 180,000 years. It then builds ten copies of itself, and sends them off to other planets, each a further ten light years-away. In this way it would take only 5,000 probe generations to fill the entire galaxy—an accomplishment that would be achieved in less than a billion years.

In addition to fulfilling a fundamental human need, our space programs have provided us with countless innovations in everyday life. As the saying goes, necessity is the mother of invention. From cell phone cameras to artificial limbs, our need to create innovative solutions for survival in harsh space has provided benefits found everywhere in society. Human nature is at its best when pushed to the creative limits.

Life on Earth and our very civilizations are indeed faced with many problems. These issues will take many years and societal changes to fix, if ever. However, it is our obligation to muster forward and remember that we are all one species on a shared planet. Our fate rests in our unity and our determination to discover the unknown. Just like those early humans who decided to sail into the unknown, we have no assurances we will find something on the other side of our ventures into space. Also like those early humans, there are some today who would describe sending people into outer space reckless.

Ultimately, what space exploration contributes to humanity is hope. Hope that when we set aside our differences and work on what can seem insurmountable, we can achieve the impossible.

"Is Space Exploration a Waste of Money?" by Sarah Treadwell, Blue Marble Space Institute of Science, August 6, 2020.

But it's not hard to imagine that an advanced civilisation might produce space probes that could travel significantly faster than ours currently do, so colonising the galaxy in just a few hundred million years is not unlikely.

But here's the thing: the Milky Way has existed for around ten billion years, and we know that some planets exist around stars that are almost this old. So if intelligent life really is common, the likelihood is that it evolved elsewhere to our stage of intelligence several billion years ago, giving it plenty of time to colonise the galaxy. So where is everybody?

Are We All Alone ...?

Entire books have been written exploring the various solutions to the Fermi paradox, but they fall into the following general categories.

Rare Earth: It may be that there are no civilisations in the galaxy any more advanced than we are. Perhaps the combination of astronomical, geological, chemical and biological factors needed to allow the emergence of complex, multicellular life is just so unlikely that it's only happened once.

Doomsday: Perhaps life and civilisations emerge often, but it is the nature of "intelligent" life to destroy itself within a few hundred years. The human race certainly has no shortage of ways of accomplishing this, whether it's via physical, chemical or biological weapons of mass destruction, or as a result of climate change, or even a nanotechnology catastrophe. If life doesn't persist very long on any planet, we shouldn't expect to see much evidence of it around the galaxy.

Extinction: Even if we don't wipe ourselves out, perhaps the universe conspires to eliminate civilisations on a regular basis? It's clear on Earth that there have been at least five mass extinctions. Some of these may have been triggered by the impact of massive asteroids, but other possible extinction causing events might include nearby supernovae or gamma-ray bursts.

...Or Are the Aliens Just Hiding?

There is another class of possible solutions to the Fermi paradox that boil down to the fact that alien civilisations do exist, but we simply see no evidence of them.

Distance scales: Perhaps civilisations are spread too thinly throughout the Galaxy to effectively communicate with each other? Civilisations may be separated in space, and also in time, so two civilisations just don't overlap during the time that they're each active.

Technical problems: Maybe we're not looking in the right place, or in the right way? Or maybe we just haven't been looking for long enough? Perhaps we've not recognised a signal that's out there, because the alien civilisation is using technology that we simply cannot comprehend.

Isolationist: Perhaps the aliens are out there, but they're choosing to hide themselves from us? Perhaps everyone is listening, but nobody is transmitting? It may be that other civilisations know we're here, but the Earth is purposely isolated, as if we're some kind of exhibit in a zoo.

Finally, there are of course the more extreme possibilities such as that the Galaxy that we observe to be empty of life is a simulation, constructed by aliens. Or perhaps the aliens are already here among us. Such speculation is great for science fiction, but without evidence, it's not worth pursuing further.

My own hunch is that life is indeed common in the galaxy, but intelligent life is rare—either because it doesn't evolve very often, or it doesn't last very long once it does. For that reason I think that SETI programmes are probably doomed to fail—although I would love to be proved wrong.

Instead I think the best chance of finding life elsewhere in the galaxy is through spectroscopy of the atmospheres of transiting terrestrial planets. That will be carried out by missions such as such as the European Space Agency's PLATO spacecraft, due for launch in 2024. Such life may just be a green slime that we can scrape off a rock with our finger, but its detection would truly transform our view of the universe, and ourselves.

"As Earth's empty spaces are filled...
capitalistkind emerges to rescue
capitalism from its terrestrial
limitations, launching space
rockets, placing satellites into orbit,
appropriating extraterrestrial
resources, and, perhaps one day,
building colonies on distant planets
like Mars."

The New Era of Space Exploration Is Only Concerned with Commercialization

Victor L. Shammas and Tomas B. Holen

In the following viewpoint, Victor L. Shammas and Tomas B. Holen argue that outer space is quickly becoming commercialized and profit-driven, thanks to a new era of space exploration by private corporations. Whereas government space programs set out to benefit humankind, the endeavors of wealthy entrepreneurs like Elon Musk, Richard Branson, and Jeff Bezos are decidedly aimed at conquering and expanding wealth. Victor L. Shammas is associate professor of sociology at the University of Agder, Norway. Tomas B. Holen is an independent scholar based in Norway.

"One Giant Leap for Capitalistkind: Private Enterprise in Outer Space," by V. L. Shammas and T. B. Holen, *Palgrave Communications* 5, 10 (2019), https://doi.org/10.1057/s41599 -019-0218-9. Licensed under CC BY 4.0 International.

As you read, consider the following questions:

1. How did SpaceX's Falcon Heavy convey the idea of a shift to capitalism, according to the viewpoint?
2. What do the authors mean by "charismatic accumulation"?
3. How have spatial fixes served to conserve the capitalist system?

On 6 February 2018, the California-based Space Exploration Technologies Corp., also known as SpaceX, launched its first Falcon Heavy rocket, a powerful, partially reusable launch vehicle, into space from Cape Canaveral Launch Complex 39 in Florida. With its significant thrust and payload capacity, the Falcon Heavy had the "ability to lift into orbit nearly 64 metric tons...a mass greater than a 737 jetliner loaded with passengers, crew, luggage and fuel" (SpaceX, 2018). Multiple reusable parts, including first-stage boosters (and, in later versions, composite payload fairing) provided a lift capacity nearly twice that of the next-most powerful rocket in operation, the United Launch Alliance's (ULA) Delta IV Heavy, and at nearly one-third the cost. With this first Falcon Heavy test flight, which produced widespread public enthusiasm and outpourings of support from both politicians and industry observers, SpaceX demonstrated that private corporations were busy redefining the domain of space exploration. SpaceX seemed to usher in an era differing markedly from that other period of astronautical excitement, the Cold War-era space race between the United States and the Soviet Union. Additionally, visions once restricted to the domain of science fiction now seemed increasingly attainable, freed from the (alleged) impediments of slow-moving nation-states: with the ascendancy of private corporations like SpaceX, satellite launches, space tourism, asteroid mining, and even the colonization of Mars seemed increasingly achievable (Cohen, 2017; Dickens and Ormrod, 2007a, 2007b; Klinger, 2017; Lewis, 1996).

In this sense, SpaceX's Falcon Heavy also carried a crucial ideological payload: the very idea of private enterprise and capitalist relations overtaking outer space. The Falcon Heavy conveyed this idea quite concretely. Onboard the rocket was an electric car, a Tesla Roadster (said to be Elon Musk's personal vehicle), which functioned as the rocket's "dummy load," playing David Bowie's "Space Oddity" and "Life on Mars?" on repeat on the car's stereo system. An enticing marketing stunt viewed by millions online through SpaceX's YouTube live stream—with 2.3 million concurrent views, it was the second biggest live stream in YouTube history (Singleton, 2018)—the Falcon Heavy test flight embraced the logic of "cool capitalism" (Schleusener, 2014), with in-jokes referencing Douglas Adams's *Hitchhiker's Guide to the Galaxy*, while heralding the arrival of a commercialized space age, dubbed by industry insiders as the age of "NewSpace."

But how are we to understand NewSpace? In some ways, NewSpace signals the emergence of capitalism in space. The production of carrier rockets, placement of satellites into orbit around Earth, and the exploration, exploitation, or colonization of outer space (including planets, asteroids, and other celestial objects), will not be the work of humankind as such, a pure species-being (Gattungswesen), but of particular capitalist entrepreneurs who stand in for and represent humanity. Crucially, they will do so in ways modulated by the exigencies of capital accumulation. These enterprising capitalists are forging a new political-economic regime in space, a post-Fordism in space aimed at profit maximization and the apparent minimization of government interference. A new breed of charismatic, starry-eyed entrepreneurs, including Musk's SpaceX, Richard Branson's Virgin Galactic, and Amazon billionaire Jeff Bezos's Blue Origin, to name but a selection, aim at becoming "capitalists in space" (Parker, 2009) or space capitalists. Neil Armstrong's famous statement will have to be reformulated: space will not be the site of "one giant leap for mankind," but rather one giant leap for capitalistkind. With the ascendancy of NewSpace, humanity's future in space will not be "ours," benefiting humanity tout court, but will rather be the result of particular capitalists, or capitalistkind, toiling to recuperate space and bring its vast domain into the fold of capital accumulation: NewSpace sees outer space as the domain of private enterprise, set to become the "first-trillion dollar industry," according to some estimates, and likely to produce the world's first trillionaires (see, e.g., Honan, 2018)—as opposed to Old Space, a derisive moniker coined by enthusiastic proponents of capitalism-in-space, widely seen to have been the sole preserve of the state and a handful of giant aerospace corporations, including Boeing and Lockheed Martin, in the Cold War-era Space Age.

Under Donald Trump's presidency, the adherents of NewSpace have found a ready political partner. The commercialization of outer space was already well under way with Obama's 2010 National Space Policy, which emphasized "promoting and supporting a competitive US commercial space sector," which was "considered vital to…continued progress in space" (Tronchetti, 2013, p. 67–68). But the Trump administration has aggressively pursued the deregulation of outer space in the service of profit margins. Wilbur Ross, President Trump's Secretary of Commerce, has eagerly supported the private space industry by pushing the dismantling of regulatory frameworks. As Ross emphatically stated, "The rate of regulatory change must accelerate until it can match the rate of technological change!" (Foust, 2018a). Trump has proposed privatizing the provision of supplies to the International Space Station (ISS) while re-establishing the Cold War-era National Space Council, which includes members from Lockheed Martin, Boeing, ULA, and a series of NewSpace actors, such as SpaceX and Blue Origin. Ross was visibly enthusiastic about SpaceX's Falcon Heavy launch in February 2018 and seemed to embrace Musk's marketing ploy. "It was really quite an amazing thing," Ross said. "At the end of it, you have that little red Tesla hurdling [sic] off to an orbit around the sun and the moon" (Bryan, 2018). That same month, Ross spoke before the National Space Council, commenting appreciatively that "space is already a $330 billion industry" that was set to become a "multitrillion-dollar one in coming decades." He noted that private corporations needed "all the help we can give them" and said it was "time to unshackle business activity in space" (Department of Commerce, 2018).

Secretary Ross's remarks followed on the heels of the American Space Commerce Free Enterprise Act, a US House of Representatives bill introduced in 2017, which, in a remarkable volte-face, unilaterally declared that "space is not a global commons," a crucial departure from ratified international treaties that paved the way for private property rights and the exploitation of precious resources in outer space. In case anyone had missed this little-noticed policy démarche, tucked away in the midst of an obscure piece of legislation, one of Trump's supporters, the executive director of the National Space Council, Scott Pace, publicly reiterated that "outer space is not...the 'common heritage of mankind,' not 'res communis,' nor is it a public good" (Pace, 2017). Instead, outer space was quickly being recast as a private good or a space for private property. As the United States became "open for business in space" (Smith, 2017), in the words of one Republican congressman, space itself was being opened up to the interests of private enterprise. The Outer Space Treaty of 1967 established space as terra nullius. One of the treaty's premises is that no celestial body can be claimed as the property of any particular state, so that "outer space...is not subject to national appropriation by claim of sovereignty, by means of use or occupation, or by any other means." While this does not prevent nations from extracting resources from celestial bodies, there is a clear requirement that these activities benefit all of Earth's inhabitants (Tronchetti, 2013, p. 14; Lyall and Larsen, 2009), paving the way for kind of communism in space which precludes the proclivities of capitalistkind. As noted, however, the Outer Space Treaty's assertion of space as a commons has come under pressure in recent years, at first in the form of so many quasi-comical ventures, bordering on fraudulent shams, with a flourishing online trade in "lunar property"— Everybody Is Saying It...Nothing Could Be Greater Than to Own Your Own Crater!—including the production of seemingly authentic land deeds that remain practically unenforceable and contravened by treaty obligations anyway. More recently, its status as commons has been denied by President Trump and leading US Republicans. Communism in space was a possibility only so long as space was materially inaccessible to capitalistkind: as space becomes a probable site of profitable ventures, the Outer Space Treaty's proto-communism must falter and fade away.

Certain parallels exist between the exploration and colonization of outer space and similar maritime ventures back on Earth. To take but one limited aspect of the overlapping legal issues raised by these two areas, that of resource exploitation: the 1982 United Nations Convention on the Law of the Sea (UNCLOS) established that the "seabed and ocean floor" beyond a nation's territorial waters (or "the Area") are the "common heritage of mankind, the exploration and exploitation of which shall be carried out for the benefit of mankind as a whole." Like outer space, Earth's seabed is part of the commons. Similarly, the International Seabed Authority, which was established to oversee the 1982 convention, is to "provide for the equitable sharing of financial and other economic benefits derived from activities in the Area" (UN, 1982, p. 71). In principle, then, any profits arising from, e.g., the mining of polymetallic nodules, are to be shared with all of humankind, including "developing States, particularly the least developed and the land-locked among them" (UN, 1982, p. 56). Whether this is likely is to happen is, according to a recent review, likely to be hampered by two factors. First, the commercial exploitation of seabed metals, which is first and foremost a technical issue, "seems as far away as ever" (Wood, 2008). Second, and perhaps more importantly, the political climate surrounding the creation and ratification (with the exception of the United States) of the 1982 convention has now appreciably shifted: "Much of the ideological passion that characterized the debates in the First Committee of the Third UN Conference on the Law of the Sea, and to some degree also in the Preparatory Commission, have now subsided" (Wood, 2008). As with outer space, the ocean floor becomes a legal site of contestation the moment states and corporations are technically capable of exploiting it.

This article adopts an approach broadly derived from the critical theory tradition to analyze NewSpace. Drawing on David Harvey's notion of spatial fixes, as well as key theoretical insights from such varied thinkers as Hegel, Marx, Bourdieu, and Deleuze and Guattari, this article asks in what ways the NewSpace paradigm can be rethought through a critical (neo-Marxist) political economy framework. Below, we advance three crucial arguments. First, there is an expedient conflation of capitalist interests with a universalizing notion of the interests of humanity. Second, the state continues to play an important role in supporting capital accumulation in space; a key tension in this area is the question of the continued role of the state in facilitating and financing NewSpace ventures—a role that is simultaneously downplayed and even, on occasion, dismissed by NewSpace actors themselves. Finally, we reassess the commercialization of space through Harvey's concept of the spatial fix, arguing that outer space serves as an important outlet for surplus capital, a site of knowledge production and technological innovation, and a potential reservoir of untapped raw materials. While the future is inherently uncertain, the article spotlights the expansive tendencies of global capital and describes the ways NewSpace actors themselves have come to view outer space as the probable future site of a post-terrestrial form of capital accumulation.

The Universalization of Capitalism

The 2010s may very well be remembered as the "Age of NewSpace," the decade when outer space was turned into a capitalist space, when private corporations pushed the price of launches, satellites, and space infrastructure downwards, exerting what industry insiders call the "SpaceX effect" (Henry, 2018), centered on the technological achievement of "reusability," recovering used rocket boosters for additional launches, promising to drastically reduce the price of going to space (Morring, 2016). As one report observes, "Not only has the number of private companies engaged in space exploration grown remarkably in recent years, these companies are quickly besting their government-sponsored competitors" (Houser, 2017). What the rockets, shuttles, ships, and landing pods will carry beneath their payload fairing or in their cargo hold, however, along with supplies and satellites, is the capitalist worldview, a particular ideology—just as Robinson Crusoe, in Marx's ironic retelling in *Capital*, "having saved a watch, ledger, ink and pen from the shipwreck... soon begins, like a good Englishman, to keep a set of books" (Marx, 1976, p. 170), brings with him English political economy—"Freedom, Equality, Property and Bentham," as Marx (1976, p. 280) says elsewhere—to his desert island.

In early 2018, astronomers across the world learned that a New Zealand start-up, Rocket Lab, which aimed to launch thousands of miniature satellites into orbit around Earth (so-called "smallsats"), had planned to launch a giant, shining "disco ball"—the "Humanity Star"—into orbit around Earth. It was an elaborate marketing stunt masked by humanistic idealism. "No matter where you are in the world, or what is happening in your life," said Rocket Lab CEO Peter Beck, "everyone will be able to see the Humanity Star in the night sky" (Amos, 2018). Many astronomers expressed outrage at these plans, fearing that the light from the Humanity Star would threaten their ability to carry out scientific observations. But while these astronomers were incensed by the idea of a bright geodesic object disrupting their ability to carry out observations, concerns with the effects of the arrival of capitalistkind on their ability to collect data were non-existent. The astronomical community was angered by the idea of a material, concrete, visible object polluting "pure" scientific data, but it paid less attention to the (invisible and abstract) recuperation of the night sky as it was brought into the fold of capitalism.

In an interview, Beck was quizzed about the Humanity Star and asked by a reporter about the difficulties of generating profits in space (Tucker, 2018). To this Beck replied, "It has always been a government domain, but we're witnessing the democratization of it...[I]t [is] turning into a commercially dominated domain." Beck established an equivalence established between the dissolution of space as the rightful domain of states and the advent of profit-making ventures as signs of "democratization." In space, according to Beck's logic, democratization involves the disappearance of the state and the rise of capital. The argument, of course, is impeccably post-statist: on this account, states are monolithic, conservative Leviathans beyond the reach of popular control; corporations, on the other hand, are in principle representatives of the everyman: in the age of the start-up, any humble citizen could in theory become an agent of disruption, a force for change, an explorer of space, and a potential member of the cadre of capitalistkind. Following this logic, the question for the entrepreneurs of NewSpace is how to monetize outer space, which means turning space into a space for capital; their question is how they can deplanetarize capital and universalize it, literally speaking, that is, turn the Universe into a universe for capital. In this light, Peter Beck's distortion of democratic ideals appears eminently sensible, equating democratization with monetization, that is, capital liberated from its earthly tethers.

Emblematic of this capitalist turn in space was the founding of Moon Express in 2011, composed of a "team of prominent Silicon Valley entrepreneurs...shooting for the moon with a new private venture aimed at scouring the lunar surface for precious metals and rare metallic elements" (Hennigan, 2011). Following Google's Lunar XPRIZE—an intertwining of Silicon Valley and NewSpace's capitalistkind—which promised a $20 million prize for the first private company to land a spacecraft on the Moon, travel 500 meters, and transmit high-definition images back to Earth, all by March 2018, Moon Express claimed that it would be capable of landing on the lunar surface and earn the cash prize. Their stated goal was twofold: first, to mine rare resource like Helium-3 (a steadily dwindling scarce resources on Earth), gold, platinum group metals, and water, and, second, to carry out scientific work that would "help researchers develop human space colonies for future generations" (Ioannou, 2017). The ordering is telling: first profits, then humanity. These were the hollow, insubstantial promises of a venture-capitalized NewSpace enterprise: in early 2018, Google announced that none of the five teams competing for the Lunar XPRIZE, including Moon Express, would reach their stated objectives by the 31 March deadline and they were taking their money back (Grush, 2018). In this sense, it was typical for NewSpace in its formative years: a corporate field populated by (overly exuberant) private enterprises who promised more than they could deliver. But the belief in NewSpace is real enough. In a tome bursting with the optimism of NewSpace, Wohlforth and Hendrix claim that "the commercial spaceflight industry is transforming our sense of possibility. Using Silicon Valley's money and innovative confidence, it will soon bring mass space products to the market" (2016, p. 7).

The trope of humanity plays a key role in the rhetoric of the adherents of NewSpace. To fulfill the objectives of NewSpace, including profit maximization and the exploitation of celestial bodies, the symbolic figure of a shared humanity serves a useful purpose, camouflaging the conquest of space by capitalism with a dream of humanity boldly venturing forth into the dark unknown, thereby also providing the legitimacy and enthusiasm needed to support bolster the legitimacy of NewSpace. So long as the stargazers and SpaceX watchers are permitted their fill of "collective effervescence," to use Durkheim's (1995, p. 228) concept, capitalist entrepreneurs will be able to pursue their business interests more or less as they please. The spectacle of outer space is crucial in this regard.

Crucially, however, and despite this spectacle, SpaceX's technology might not necessarily be more sophisticated than its competitors or predecessors. Some industry insiders have rebuffed some of the more the spectacular claims of NewSpace's proponents, arguing that launch vehicle reusability requires a (perhaps prohibitively) expensive refurbishing of the rocket engines involved in launches: "The economics will depend on how many times a booster can be flown, and how much the individual expense will be to refurbish the booster…each time" (Chang, 2017). Reusability may be a technological dead-end because of the inherently stressful effects of a rocket launch on the launch vehicle's components, with extreme limitations on reusability beyond second-use as well as added risks of malfunctions that customers and insurers are likely to wish to avoid. Furthermore, the Falcon Heavy still has not matched the power and payload capacity of NASA's Saturn V, a product of 1960s military-industrial engineering and Fordist state spending programs. What SpaceX and other NewSpace corporations do with great ingenuity, however, is to manage the spectacle of outer space, producing outpourings of public fervor, aided by a widespread adherence to the "Californian Ideology" (Barbrook and Cameron, 1996), or post-statist techno-utopianism, in many postindustrialized societies.

The very centrality of these maneuvers has initiated a new phase in the history of capitalist relations, that of "charismatic accumulation"—certainly not in the sense of any "objective" or inherent charismatic authority, but with a form of illusio, to speak with Bourdieu, vested in the members of capitalistkind by their uncanny ability to spin mythologizing self-narratives. This has always been part of the capitalist game, from Henry Ford and onwards, but the charismatic mission gains a special potency in the grandiose designs of NewSpace's entrepreneurs. Every SpaceX launch is a quasi-religious spectacle, observed by millions capable of producing a real sense of wonder in a condition of (legitimizing) collective effervescence.

Outer space necessarily reduces inter-human difference to a common denominator or a shared species-being. An important leitmotiv in many Hollywood science fiction movies, including *Arrival* (2016), is that a first encounter with an alien species of intelligent beings tends to flatten all human difference (including ethnoracial and national categories), thereby restoring humankind to its proper universality (see also Novoa, 2016). Ambassadors of Earth as a whole, not representatives of particular nations, step forth to meet alien emissaries. But even in the absence of such an encounter, the search for habitable domains (or rather, profitable locales) beyond Earth will necessarily forge a shared conception of the human condition, initiated with the Pale Blue Dot photograph in 1990. Typical of this sentiment are the words of the astronomer Carl Sagan, who famously observed of this photograph: "On it everyone you love, everyone you know, everyone you ever heard of, every human being who ever was, lived out their lives."

This naïvely humanistic vision has been one of the dominant tropes in the discourse on space since the 1950s, and it remains strong today, as with the claims of the United Nations Office for Outer Space Affairs (UNOOSA) that their task is to "uphold the vision of a more equitable future for all humankind through shared achievements in space." This representational tendency mobilizes humanism to generate enthusiasm about space-related activities. But such representations are increasingly being recuperated by capitalist enterprise, so that it is not humankind but its modulation by space capitalists that will launch into the dark unknown. It is not humankind but capitalistkind that ventures forth. In early 2018, NASA was set to request $150 million in its 2019 budget to "enable the development and maturation of commercial entities and capabilities which will ensure that commercial successors to the ISS...are operational when they are needed," only one of many signs that space is becoming a space for capitalism. According to one estimate, the value of just one single asteroid would be more than $20 trillion in rare earth and platinum-group metals (Lewis, 1996), a precious prize indeed for profit-hungry corporations. Even the UNOOSA spoke vociferously in favor of the commercialization of space, appealing variously to the "industry and private sector" and elevating the "space economy" to a central pillar in its Space2030 Agenda (including the "use of resources that create and provide value and benefits to the world population in the course of exploring, understanding and utilizing space"), even as the UN agency falls back on a humanistic, almost social-democratic vision of the equitable distribution of benefits (and profits) from space mining, exploration, and colonization (UNOOSA, 2018).

We find evidence of this strategic humanism in all manner of pronouncements from NewSpace entrepreneurs. To take but one example: Naveen Jain, the chairman and co-founder of MoonEx, a lunar commercialization firm, has claimed that "from an entrepreneur's perspective, the moon has never truly been explored." The moon, Jain has claimed, "could hold resources that benefit Earth and all humanity" (Hennigan, 2011). We should note the recourse to the trope of all of humanity by this NewSpace entrepreneur, mimicked in the 1979 Moon Agreement, a UN treaty, which also held that the Moon's resources are "the common heritage of mankind" (Tronchetti, 2013, p. 13). In a purely factual sense, of course, Jain is wrong: Google Moon offers high-resolution images of the lunar surface, and the moon has already been explored, in the sense of being mapped, albeit rudimentarily and with room for further data collection. Crucially, however, these cartographic techniques have not been put to capitalist uses: mapping minerals, for instance, or producing detailed schemata that might one day turn the Moon into a "gas station" for commercial space ventures, as Wilbur Ross, Trump's Secretary of Commerce, has proposed (Bryan, 2018). What is lacking, in short, are capitalist maps of the Moon, i.e., a cartography for capital. But as Klinger (2017: 199) notes, even though no one is "actively mining the Moon" at present, at least "six national space programs, fifty private firms, and one graduate engineering program, are intent on figuring out how to do so"; furthermore, Klinger draws attention to mapping efforts that have revealed high an abundance of rare earth metals, thorium, and iron in the Moon's "Mare Procellarum KREEP" region (Klinger, 2017, p. 203).

We have already noted that it is not humanity, conceived as species-being, a Gattungswesen, that makes its way into space. The term Gattungswesen, of course, has a long intellectual pedigree, harking back to Hegel, Feuerbach, Marx, and others. The term can "be naturally applied both to the individual human being and to the common nature or essence which resides in every individual man and woman," Allan Wood (2004, p. 17) writes, as well as "to the entire human race, referring to humanity as a single collective entity or else to the essential property which characterizes this entity and makes it a single distinctive thing in its own right." Significantly, the adherents of NewSpace often resort to the idea of humanity in its broad universality (e.g., Musk, 2017), but this denies and distorts the modulation of humanity by its imbrication with the project of global (and post-global, i.e., space-bound) capitalism. It is precisely the sort of false universality implied in the humanism of the supporters of NewSpace that Marx subjected to a scathing critique in the sixth of his Theses on Feuerbach. Here Marx noted that the human essence is not made up of some "abstraction inherent in each single individual" (1998, p. 570). Instead, humans are defined by the "ensemble of social relations" in which they are enmeshed. Under NewSpace, it is not humanity, plain and simple, that ventures forth, but a specific set of capitalist entrepreneurs, carrying a particular ideological payload, alongside their satellites, instruments, and supplies, a point noted by other sociologists of outer space, or "astrosociologists" (Dickens and Ormrod, 2007a, 2007b).

The Spatial Fix of Outer Space

No longer terra nullius, space is now the new terra firma of capitalistkind: its naturalized terroir, its next necessary terrain. The logic of capitalism dictates that capital should seek to expand outwards into the vastness of space, a point recognized by a recent ethnography of NewSpace actors (Valentine, 2016, p. 1050). The operations of capitalistkind serve to resolve a series of (potential) crises of capitalism, revolving around the slow, steady decline of spatial fixes (see e.g., Harvey, 1985, p. 51–66) as they come crashing up against the quickly vanishing blank spaces remaining on earthly maps and declining (terrestrial) opportunities for profitable investment of surplus capital (Dickens and Ormrod, 2007a, p. 49–78).

A "spatial fix" involves the geographic modulation of capital accumulation, consisting in the outward expansion of capital onto new geographic terrains, or into new spaces, with the aim of filling a gap in the home terrains of capital. Jessop (2006, p. 149) notes that spatial fixes may involve a number of strategies, including the creation of new markets within the capitalist world, engaging in trade with non-capitalist economies, and exporting surplus capital to undeveloped or underdeveloped regions. The first two address the problem of insufficient demand and the latter option creates a productive (or valorizing) outlet for excess capital. Capitalism must regularly discover, develop, and appropriate such new spaces because of its inherent tendency to generate surplus capital, i.e., capital bereft of profitable purpose. In Harvey's (2006, p. xviii) terms, a spatial fix revolves around "geographical expansions and restructuring…as a temporary solution to crises understood…in terms of the overaccumulation of capital." It is a temporary solution because these newly appropriated spaces will in turn become exhausted of profitable potential and are likely to produce their own stocks of surplus capital; while "capital surpluses that otherwise stood to be devalued, could be absorbed through geographical expansions and spatiotemporal displacements" (Harvey, 2006, p. xviii), this outwards drive of capitalism is inherently limitless: there is no end point or final destination for capitalism. Instead, capitalism must continuously propel itself onwards in search of pristine sites of renewed capital accumulation. In this way, Harvey writes, society constantly "creates fresh productive powers elsewhere to absorb its overaccumulated capital" (Harvey, 1981, p. 8).

Historically, spatial fixes have played an important role in conserving the capitalist system. As Jessop (2006, p. 149) points out, "The export of surplus money capital, surplus commodities, and/or surplus labour-power outside the space(s) where they originate enabled capital to avoid, at least for a period, the threat of devaluation." But these new spaces for capital are not necessarily limited to physical terrains, as with colonial expansion in the nineteenth century; as Greene and Joseph (2015) note, various digital spaces, such as the Internet, can also be considered as spatial fixes: the Web absorbs overaccumulated capital, heightens consumption of virtual and physical goods, and makes inexpensive, flexible sources of labor available to employers. Greene and Joseph offer the example of online high-speed frequency trading as a digital spatial fix that furthers the "annihilation of space by time" first noted by Marx in his *Grundrisse* (see Marx, 1973, p. 524).

Outer space serves at least two purposes in this regard. In the short-to medium-term, it allows for the export of surplus capital into emerging industries, such as satellite imaging and communication. These are significant sites of capital accumulation: global revenues in the worldwide satellite market in 2016 amounted to $260 billion (SIA, 2017, p. 4). Clearly, much of this activity is taking place "on the ground"; it is occurring in the "terrestrial economy." But all that capital would have to find some other meaningful or productive outlet were it not for the expansion of capital into space. Second, outer space serves as an arena of technological innovation, which feeds back into the terrestrial economy, helping to avert crisis by pushing capital out of technological stagnation and innovation shortfalls.

In short, outer space serves as a spatial fix. It swallows up surplus capital, promising to deliver valuable resources, technological innovations, and communication services to capitalists back on Earth. This places outer space on the same level as traditional colonization, analyzed in Hegel's *Philosophy of Right*, which Hegel thought of as a product of the "inner dialectic of civil society," which drives the market to "push beyond its own limits and seek markets, and so its necessary means of subsistence, in other lands which are either deficient in the goods it has overproduced, or else generally backward in creative industry, etc." (Hegel, 2008, p. 222). In this regard, SpaceX and related ventures are not so very different from maritime colonialists and the trader-exploiters of the British East India Company. But there is something new at stake. As the Silicon Valley entrepreneur Peter Diamandis has gleefully noted: "There are twenty-trillion-dollar checks up there, waiting to be cashed!" (Seaney and Glendenning, 2016). Capitalistkind consists in the naturalization of capitalist consciousness and practice, the (false) universalization of a particular mode of political economy as inherent to the human condition, followed by the projection of this naturalized universality into space—capitalist humanity as a Fukuyamite "end of history," the end-point of (earthly) historical unfolding, but the starting point of humanity's first serious advances in space.

What role, then, for the state? The frontiersmen of NewSpace tend to think of themselves as libertarians, pioneers beyond the domain of state bureaucracy (see Nelson and Block, 2018). "The government should leave the design work and ownership of the product to the private sector," the author of a 2017 report, Capitalism in Space, advocates. "The private companies know best how to build their own products to maximize performance while lowering cost" (Zimmerman, 2017, p. 27). One ethnographer notes that "politically, right-libertarianism prevails" amongst NewSpace entrepreneurs (Valentine, 2016, p. 1047–1048). Just as Donald Rumsfeld dismissed the opponents to the Iraq War as "Old Europe," so too are state entities' interests in space exploration shrugged off as symptoms of "Old Space." Elon Musk, we are told in a recent biography, unlike the sluggish Big State actors of yore, "would apply some of the start-up techniques he'd learned in Silicon Valley to run SpaceX lean and fast...As a private company, SpaceX would also avoid the waste and cost overruns associated with government contractors" (Vance, 2015, p. 114). This libertarianism-in-space has found a willing chorus of academic supporters. The legal scholar Virgiliu Pop introduces the notion of the frontier paradigm (combining laissez-faire economics, market competition, and an individualist ethic) into the domain of space law, claiming that this paradigm has "proven its worth on our planet" and will "most likely...do so in the extraterrestrial realms" as well (Pop, 2009, p. vi). This frontier paradigm is not entirely new: a "Columbus mythology," centering on the "noble explorer," was continuously evoked in the United States during the Cold War space race (Dickens and Ormrod, 2016, pp. 79, 162–164).

But the entrepreneurial libertarianism of capitalistkind is undermined by the reliance of the entire NewSpace complex on extensive support from the state, "a public-private financing model underpinning long-shot start-ups" that in the case of Musk's three main companies (SpaceX, SolarCity Corp., and Tesla) has been underpinned by $4.9 billion in government subsidies (Hirsch, 2015). In the nascent field of space tourism, Cohen (2017) argues that what began as an almost entirely private venture quickly ground to a halt in the face of insurmountable technical and financial obstacles, only solved by piggybacking on large state-run projects, such as selling trips to the International Space Station, against the objections of NASA scientists. The business model of NewSpace depends on the taxpayer's dollar while making pretensions to individual self-reliance. The vast majority of present-day clients of private aerospace corporations are government clients, usually military in origin. Furthermore, the bulk of rocket launches in the United

States take place on government property, usually operated by the US Air Force or NASA.

This inward tension between state dependency and capitalist autonomy is itself a product of neoliberalism's contradictory demand for a minimal, "slim" state, while simultaneously (and in fact) relying on a state reengineered and retooled for the purposes of capital accumulation (Wacquant, 2012). As Lazzarato writes, "To be able to be 'laissez-faire,' it is necessary to intervene a great deal" (2017, p. 7). Space libertarianism is libertarian in name only: behind every NewSpace venture looms a thick web of government spending programs, regulatory agencies, public infrastructure, and universities bolstered by research grants from the state. SpaceX would not exist were it not for state-sponsored contracts of satellite launches. Similarly, in 2018, the US Defense Advanced Research Projects Agency (DARPA)—the famed origin of the World Wide Web—announced that it would launch a "responsive launch competition," meaning essentially the reuse of launch vehicles, representing an attempt by the state to "harness growing commercial capabilities" and place them in the service of the state's interest in ensuring "national security" (Foust, 2018b).

This libertarianism has been steadily growing in the nexus between Silicon Valley, Stanford University, Wall Street, and the Washington political establishment, which tend to place a high value on Randian "objectivism" and participate in a long American intellectual heritage of individualistic "bootstrapping" and (allegedly) gritty self-reliance. But as Nelson and Block (2018, p. 189–197) recognize, one of the central symbolic operations of capitalistkind resides in concealing its reliance on the state by mobilizing the charm of its entrepreneurial constituents and the spectacle of space. There is a case to be made for the idea that SpaceX and its ilk resemble semi-private corporations like the British East India Company. The latter, "incorporated by royal charter from Her Majesty Queen Elizabeth I in 1600 to trade in silk and spices, and other profitable Indian commodities," recruited soldiers and built a "commercial business [that] quickly became a business of conquest" (Tharoor, 2017). SpaceX, too, is increasingly imbricated with an attempt on the part of a particular state, the United States, to colonize and appropriate resources derived from a particular area, that of outer space; it, too, depends on the infrastructure, contracts, and regulatory environment that thus far only a state seems able to provide. Its private character, like that of the East India Company, is troubled by being deeply embedded in the state. As one commentator has observed of SpaceX, "If there's a consistent charge against Elon Musk and his high-flying companies…it's that they're not really examples of independent, innovative market capitalism. Rather, they're government contractors, dependent on taxpayer money to stay afloat" (cit. Nelson and Block, 2018, p. 189).

Perhaps this should not come as a surprise. As Bourdieu (2005, p. 12) observed, "The economic field is, more than any other, inhabited by the state, which contributes at every moment to its existence and persistence, and also to the structure of the relations of force that characterize it." The state lays out the preconditions for market exchanges. Under neoliberalism, the state is the preeminent facilitator of markets. The neoliberal state is not so much a Minimalstaat, night watchman state, or slim state as it is the prima causa of market society (see, e.g., Wacquant, 2012). Similarly, in the political theory of Deleuze and Guattari, any economic development presupposes the political differentiation caused by the state (Deleuze and Guattari, 2004a, p. 237–238). Even in the global environment of contemporary capitalism, the market cannot operate without the state becoming integrated with capitalism itself, as "it is the modern state that gives capitalism its models of realization" (Deleuze and Guattari, 2004b, p. 480). For capitalism to survive in outer space, the state must create a regulatory environment, subsidize infrastructure, and hand down contracts—in short, assemble outer space as a domain made accessible in legal, technical, and economic ways.

Universalizing Capital

As Earth's empty spaces are filled, as our planet comes to be shorn of blank places, capitalistkind emerges to rescue capitalism from its terrestrial limitations, launching space rockets, placing satellites into orbit, appropriating extraterrestrial resources, and, perhaps one day, building colonies on distant planets like Mars. But why limit ourselves to Mars? As of mid-2017, NASA's Kepler observatory had discovered more than 5000 exoplanets—planets that seem like promising alternatives to Earth, located at an appropriate distance from their respective suns in the famed "Goldilocks zone." These "planetary candidates," as they are known—that is, candidates for the replacement of Earth, capable of supporting human life with only minimal technological augmentation or cybernetic re-engineering—are above all viable candidates for selection by specific capitalists seeking to discover new profitable ventures beyond the limits of an Earth-bound capitalism. Space reveals the impotence of the neoliberal, post-Fordist state, its incapacity and unwillingness to embark on gigantic infrastructural projects, to project itself outwards, and to fire the imagination of (actual) humankind. Capitalistkind steps in to fill the vacuum left behind by a state that lacks what Mann (2012, p. 170) calls "infrastructural power." The old question, the question of Old Space, was quite simply: is this planet a viable site for humankind, a suitable homeland for the reproduction of human life away from Earth? But the new question, the question for NewSpace, will be: can this celestial body support capitalistkind? Will it support the interests of capitalist entrepreneurs, answering to the capitalist desire for continued accumulation?

While some elements of the astrosociological community, such as the Astrosociology Research Institute (ARI), insist on elucidating the "human dimension" in outer space, Dickens and Ormrod recognize that this humanization-through-capitalism really involves the "commodification of the universe" (2007b, p. 2). While Dickens and Ormrod develop similar arguments to those sketched here—from their concept of an "outer spatial fix" to their argument about outer space becoming woven into circuits of capital accumulation—they were writing at a time when their remarks necessarily remained speculative: the commercialization of space was still in its infancy. In an inversion of Hegel's owl of Minerva, reality has since largely confirmed their ideas and caught up with theory. Above all, when considering the various ventures ongoing in space today, it is not so much the universalizing human dimension as the specifically capitalist dimension that is striking. With the advent of NewSpace, outer space is becoming not the domain of a common humanity but of private capital.

The arguments laid out above mirror an ongoing turn in critical scholarship away from the notion of the Anthropocene towards a more rigorously political-economic concept of Capitalocene, premised on the "claim that capitalism is the pivot of today's biospheric crisis" (Moore, 2016, p. xi). Just as the exponents of the concept of Capitalocene emphasize that it is capitalism, and not humanity as such, that is the driving force behind environmental transformation, so too does the notion of capitalistkind emphasize that it is not humankind tout court but rather a set of specific capitalist entrepreneurs who are acting as the central transformative agents of and in outer space, with the "ever-increasing infiltration of capital" into what was formerly the domain of the state (Dickens and Ormrod, 2007a, p. 6). We can also think about these issues in terms of what Philippopoulos-Mihalopoulos (2015) terms "spatial justice." This concept captures the fact that struggles over justice are often struggles to occupy space, as the term is more conventionally understood, as with urban battles over the "right to the city" (Harvey, 2008), to provide just one example. But the same also holds true for outer space: there is an ongoing struggle over the right to take up space in outer space. So far, the capitalist side appears to be winning. As the proto-communism of the Cold War-era Outer Space Treaty is abandoned—in tandem with the increased technological feasibility of exploiting resources and accumulating profits in outer space—spatial justice in outer space increasingly comes to mean the "justice" of capital, capitalistkind taking the place of humankind. It is comparatively easy to declare that outer space is a commons, as the Outer Space Treaty did in the late 1960s, when that domain is, for all practical purposes, inaccessible to capital; with the heightened accessibility of outer space, however, it is unsurprising that central political agents, such as President Trump's administration, should seek to dismantle this regulatory framework and ensure the smooth functioning of capital accumulation beyond the terrains of Earth.

What kind of capitalism is being projected into space? The complexity of state-market relations is sufficient to force us to hedge against a simplified reading of space commercialization: it is not a matter of states against markets, as if the two were mutually exclusive. Instead, as Bratton (2015) suggests, we are witnessing the emergence of a "stack," a complex intertwining of commercial, geopolitical, and technological concerns, which challenges previous notions of state sovereignty. This can be seen as a hybridized state-market form, with technology playing a central role in reciprocal processes of political and economic transformation. On the one hand, outer space was in some sense always already the domain of marketization, albeit to a limited extent, even during the Cold War, from the first commercial satellite launch in the early 1960s to President Ronald Reagan's implementation of the Commercial Space Launch Act of 1984, which aimed to encourage private enterprise to take an interest in an emerging launch market. As Hermann Bondi, the head of the European Space Organization, wrote in the early 1970s, "It is clear…that there must be three partners in space, universities and research institutions on the one hand, the government on the second and industry on the third" (Bondi, 1971, p. 9).

On the other hand, outer space still remains firmly within the domain of the state and is likely to do so for the foreseeable future, with the likely continued importance of military uses of satellite technology and the weaponization of Earth's orbit—crucially, the Outer Space Treaty only prohibits nuclear arms and other "weapons of mass destruction" in space, not conventional weapons, such as ballistic missiles. One novel element in this phase of capitalism-in-space is the interrelationship between Silicon Valley, NewSpace, and the state (see, e.g., Vance, 2015). Silicon Valley's capitalist class, including Amazon's Jeff Bezos, play an outsize role in NewSpace. Behind and around these figures, however, remains the state—through its weighty fiscal, regulatory, military, and symbolic investments. To take but one example: In June 2018, SpaceX won a $130 million contract with the US Air Force to launch an "Air Force Space Command" satellite onboard a Falcon Heavy rocket (Erwin, 2018).

Fredric Jameson's (2003, p. 76) oft-quoted observation that it is easier to imagine the end of humankind than the end of capitalism, is realized in the ideals and operations of capitalistkind. Elon Musk has observed that the goal of SpaceX is to establish humankind as a "multiplanetary species with a self-sustaining civilization on another planet" whose purpose is to counteract the possibility of a "worst-case scenario happening and extinguishing human consciousness" (Vance, 2015, p. 5). But couldn't we view this idealistic assertion on behalf of humanity in another way? It is not human consciousness, over and against what the writer Kim Stanley Robinson (2017, p. 2) calls "mineral unconsciousness" (i.e., the mute, geological reality of the natural universe), so much as a specifically capitalist consciousness that is at stake. While the actions of capitalistkind may primarily be aimed at ensuring the future survival of the human species, an additional result is to ensure that the very idea of capitalism itself will outlive a (distantly) possible extinction event. Capitalism is a self-replicating system, pushing to expand ever outwards, using a territorializing strategy of survival. As David Harvey notes, "a steady rate of growth is essential for the health of a capitalist economic system, since it is only through growth that profits can be assured and the accumulation of capital be sustained" (1990, p. 180). In this respect, outer space is ideal: it is boundless and infinite. As Earth comes to be blanketed by capital, it is only to be expected that capital should set its sights on the stars above. The actions of capitalistkind serve to bolster the capitalist mode of production and accumulation: it is not only life but capital itself that must outlive Earth—even into the darkness of space.

> *"The range of entirely novel environments opened to investigation will be essentially limitless, and so has the potential to be a never-ending source of scientific and intellectual stimulation."*

Looking for Aliens Is Good for Society (Even If There Aren't Any)

Ian Crawford

In the following viewpoint, Ian Crawford makes the intriguing argument that the search for extraterrestrial intelligence is in and of itself a worthy goal, even if there are no aliens out there. His reasons are not quite the same as the ones mentioned in the previous viewpoints. Instead, he argues that the search could improve the culture of science and give humans a better sense of our place in the universe, a perspective that not only will lead to many scientific discoveries and advances but that also may ultimately save us from catastrophe. In fact, writes Crawford, humanity has the responsibility to do so. Ian Crawford is professor of planetary science and astrobiology at Birkbeck, University of London.

As you read, consider the following questions:

1. What are the benefits of moving science away from extreme specialization?

2. How, according to this viewpoint, might the search for alien life bring together the people of Earth (even if we never find such life)?

3. Why does humanity need what this author calls "broader perspectives" if it is to survive?

The search for life elsewhere in the universe is one of the most compelling aspects of modern science. Given its scientific importance, significant resources are devoted to this young science of astrobiology, ranging from rovers on Mars to telescopic observations of planets orbiting other stars.

The holy grail of all this activity would be the actual discovery of alien life, and such a discovery would likely have profound scientific and philosophical implications. But extraterrestrial life has not yet been discovered, and for all we know may not even exist. Fortunately, even if alien life is never discovered, all is not lost: simply searching for it will yield valuable benefits for society.

Why is this the case?

First, astrobiology is inherently multidisciplinary. To search for aliens requires a grasp of, at least, astronomy, biology, geology, and planetary science. Undergraduate courses in astrobiology need to cover elements of all these different disciplines, and postgraduate and postdoctoral astrobiology researchers likewise need to be familiar with most or all of them.

By forcing multiple scientific disciplines to interact, astrobiology is stimulating a partial reunification of the sciences. It is helping to move 21st-century science away from the extreme specialisation of today and back towards the more interdisciplinary outlook that prevailed in earlier times.

By producing broadminded scientists, familiar with multiple aspects of the natural world, the study of astrobiology therefore

THE SEARCH IS WORTH IT

My personal opinion is that yes, it is worthwhile to continue the search for extraterrestrial life, but there are many arguments both for and against this search.

Many argue that since the probability of finding evidence of intelligent life is so small, it is a waste of money to continue scanning the sky in search of extraterrestrial intelligence. However, supporters of SETI and other programs note that the amount of taxpayer money that goes into these programs is less than one military helicopter per year.

Several estimates exist as far as just how unlikely it is that we will receive a signal from extraterrestrial life. One of these uses the Drake equation, which estimates the number of technological civilizations in the galaxy. This equation takes into account the number of stars in the galaxy, fraction of those stars having planets, fraction of those planets which are habitable, fraction of those habitable planets on which life originates, fraction of that life which evolves to become intelligent, fraction of those intelligent civilizations which develop technology, and the chance that these technological civilizations are alive at the same time as us. Using generous estimates for the factors in the Drake equation, there are about a million civilizations in the galaxy, which puts the nearest one at about 150 light-years away.

It seems that searching for intelligent life is a discouraging task, due to the huge distances involved and the fact that we may never find anything at all. . . . Some people are afraid of what SETI may find, imagining malevolent aliens who may come and eat us all up . . . but I believe it is worth it to keep on looking, if for no other reason than the other discoveries that go along with it. The search for extraterrestrial life is one of the big motivations for further study of the rest of our solar system, especially Mars and the satellites of Jupiter and Saturn. Also, listening to the radio signals that arrive here from the rest of the universe has resulted in several important discoveries that may not have occurred if we were not listening for other life.

"Is It Worthwhile to Continue the Search for Extraterrestrial Life?" by Cathy Jordan, Curious Astronomer, January 28, 2019.

enriches the whole scientific enterprise. It is from this cross-fertilization of ideas that future discoveries may be expected, and such discoveries will comprise a permanent legacy of astrobiology, even if they do not include the discovery of alien life.

It is also important to recognise that astrobiology is an incredibly open-ended endeavour. Searching for life in the universe takes us from extreme environments on Earth, to the plains and sub-surface of Mars, the icy satellites of the giant planets, and on to the all-but-infinite variety of planets orbiting other stars. And this search will continue regardless of whether life is actually discovered in any of these environments or not. The range of entirely novel environments opened to investigation will be essentially limitless, and so has the potential to be a never-ending source of scientific and intellectual stimulation.

The Cosmic Perspective

Beyond the more narrowly intellectual benefits of astrobiology are a range of wider societal benefits. These arise from the kinds of perspectives—cosmic in scale—that the study of astrobiology naturally promotes.

It is simply not possible to consider searching for life on Mars, or on a planet orbiting a distant star, without moving away from the narrow Earth-centric perspectives that dominate the social and political lives of most people most of the time. Today, the Earth is faced with global challenges that can only be met by increased international cooperation. Yet around the world, nationalistic and religious ideologies are acting to fragment humanity. At such a time, the growth of a unifying cosmic perspective is potentially of enormous importance.

In the early years of the space age, the then US ambassador to the United Nations, Adlai Stevenson, said of the world: "We can never again be a squabbling band of nations before the awful majesty of outer space." Unfortunately, this perspective is yet to sink deeply into the popular consciousness. On the other hand, the wide public interest in the search for life elsewhere means that astrobiology can act as a powerful educational vehicle for the popularisation of this perspective.

Indeed, it is only by sending spacecraft out to explore the solar system, in large part for astrobiological purposes, that we can obtain images of our own planet that show it in its true cosmic setting.

In addition, astrobiology provides an important evolutionary perspective on human affairs. It demands a sense of deep, or big, history. Because of this, many undergraduate astrobiology courses begin with an overview of the history of the universe. This begins with the Big Bang and moves successively through the origin of the chemical elements, the evolution of stars, galaxies, and planetary systems, the origin of life, and evolutionary history from the first cells to complex animals such as ourselves. Deep history like this helps us locate human affairs in the vastness of time, and therefore complements the cosmic perspective provided by space exploration.

Political Implications

There is a well-known aphorism, widely attributed to the Prussian naturalist Alexander von Humboldt, to the effect that "the most dangerous worldview is the worldview of those who have not viewed the world." Humboldt was presumably thinking about the mind-broadening potential of international travel. But familiarity with the cosmic and evolutionary perspectives provided by astrobiology, powerfully reinforced by actual views of the Earth from space, can surely also act to broaden minds in such a way as to make the world less fragmented and dangerous.

I think there is an important political implication inherent in this perspective: as an intelligent technological species, that now dominates the only known inhabited planet in the universe, humanity has a responsibility to develop international social and political institutions appropriate to managing the situation in which we find ourselves.

In concluding his monumental *Outline of History* in 1925, HG Wells famously observed: "Human history becomes more and more a race between education and catastrophe." Such an observation appears especially germane to the geopolitical situation today, where apparently irrational decisions, often made by governments (and indeed by entire populations) seemingly ignorant of broader perspectives, may indeed lead our planet to catastrophe.

Periodical and Internet Sources Bibliography

The following articles have been selected to supplement the diverse views presented in this chapter.

Ella Adams, "Spending on Space Is Wasteful," *The Appalachian*, March 5, 2021. https://theappalachianonline.com/opinion -spending-on-space-is-wasteful/

Payal Dahr, "The Search for Extraterrestrial Intelligence Gets a Major Upgrade," *IEEE Spectrum*, April 17, 2020. https://spectrum.ieee .org/tech-talk/aerospace/astrophysics/search-extraterrestrial -intelligence-major-upgrade

Andrew Dempster, "Why Spending on Space Is Worth It," The Feed, May 14, 2018. https://www.sbs.com.au/news/the-feed/expert -analysis-why-spending-on-space-is-worth-it

Graham Flanagan and Dave Mosher, "The Government Stopped Funding the Search for Aliens—and This Astronomer Says That's a Big Problem," *Business Insider*, September 30, 2016. https:// www.businessinsider.com/nasa-search-intelligent-alien-life-seti -funding-2016-9

Pallab Ghosh, "Astronomers Want Public Funds for Intelligent Life Search," BBC News, February 15, 2020. https://www.bbc.com /news/science-environment-51223704

Andrew Glikson, "Before We Colonise Mars, Let's Look to Our Problems on Earth," The Conversation, December 27, 2017. https://theconversation.com/before-we-colonise-mars-lets-look -to-our-problems-on-earth-87770

Jeffrey Kluger, "Why We Should Be Spending More on Space Travel," *Time*, April 12, 2021. https://time.com/5953877/spend-more-on -space-travel/

Marina Koren, "Congress Is Quietly Nudging NASA to Look for Aliens," *The Atlantic*, May 9, 2018. https://www.theatlantic .com/science/archive/2018/05/seti-technosignatures-nasa-jill -tarter/558512/

Megan McArdle, "The Billionaires' Space Race Benefits the Rest of Us. Really," *Washington Post*, July 13, 2021. https://www

.washingtonpost.com/opinions/2021/07/13/billionaires-space
-race-benefits-rest-us-really/

Naomi Oreskes, "Exploring Space Can Unite the US—but Not in
the Way You Might Think," *Scientific American*, January 1, 2021.
https://www.scientificamerican.com/article/exploring-space-can
-unite-the-us-not-in-the-way-you-might-think/

Nicholas Reimann, "Leaving a Planet in Crisis: Here's Why Many
Say the Billionaire Space Race Is a Terrible Idea," *Forbes*, July 12,
2021. https://www.forbes.com/sites/nicholasreimann/2021/07/12
/leaving-a-planet-in-crisis-heres-why-many-say-the-billionaire
-space-race-is-a-terrible-idea/?sh=5e0f306c77c9

Nicholas Russell, "Revive the US Space Program? How About Not,"
The Guardian, April 9, 2021. https://www.theguardian.com
/commentisfree/2021/apr/09/us-space-program-nasa-earth
-problems

Matthew S. Williams, "Is It Worth It? The Costs and Benefits of Space
Exploration," Interesting Engineering, April 17, 2019. https://
interestingengineering.com/is-it-worth-it-the-costs-and-benefits
-of-space-exploration

Jason Wright, "NASA Should Start Funding SETI Again," *Scientific
American*, February 7, 2018. https://blogs.scientificamerican
.com/observations/nasa-should-start-funding-seti-again/

For Further Discussion

Chapter 1

1. One viewpoint author says that life is easy to recognize, but difficult to define. What does the author mean by that? Can you think of other things that are easy to recognize but difficult to define?
2. Several of the viewpoints in this chapter attempt to make estimates of the probability of life elsewhere in the universe (now or in the past). These estimates are based on different assumptions than previous findings, but they are still based on assumptions. Why do these assumptions matter? Why are some assumptions better than others?
3. After reading the viewpoints in this chapter, what do you think are the chances that other planets are home to civilizations similar to ours? What about less advanced life forms?

Chapter 2

1. In the first viewpoint in this chapter, the author says that he does not think that belief in "UFOs" is crazy. Then he points out that UFO belief is associated with "a tendency toward social anxiety, paranoid ideas and transient psychosis." Is he being inconsistent? Disingenuous? Why or why not?
2. Does the 2021 Department of Defense report on unidentified aerial phenomena influence or change your thinking on 'Oumuamua? Why or why not?
3. In this chapter, Simon Goodwin, himself an astrophysicist, explains why scientists are typically careful not to make extraordinary claims—even when they think they may, possibly, be true. Do you think this is a prudent approach? Why or why not?

Chapter 3

1. Do you think it is proper for a few people to decide what a message sent to potential extraterrestrial civilizations should contain, and even if it should be sent? How do you think decisions that affect the population of the entire world should be made? Do you see any parallels to the COVID-19 pandemic?

2. Do you think you might have approached the question raised in this chapter differently if you had not read the previous chapter's viewpoints? Does the fact that serious scientists take the possibility of advanced alien civilizations at least somewhat seriously make you feel more cautious about sending out signals? Why or why not?

3. If you imagine that alien life forms learn about us from our media broadcasts—internet, television, and so on—what do you imagine they think of us? Do you think they would get an accurate picture of what humans are like? If not, what might we show them that would be true, but also more accurate?

Chapter 4

1. Several of the viewpoints in this chapter discuss the difficulties SETI has obtaining funding. Why do you think the government would not want to pay for the search for extraterrestrial intelligence? Why might they want to?

2. Many experts on both sides of the debate agree that space exploration has benefits for humanity. Do you think that these benefits are worth the cost? Why or why not?

3. Much of space exploration today (including, but not limited to, the search for aliens) is funded by extremely wealthy people rather than governments. Do you think this is a good thing? Why or why not?

Organizations to Contact

The editors have compiled the following list of organizations concerned with the issues debated in this book. The descriptions are derived from materials provided by the organizations. All have publications or information available for interested readers. The list was compiled on the date of publication of the present volume; the information provided here may change. Be aware that many organizations take several weeks or longer to respond to inquiries, so allow as much time as possible.

Alternative Earth's Astrobiology Center

2460 Geology Building
Riverside, CA 92521
(951) 827-3106
email: astrobiology@ucr.edu
website: www.astrobiology.ucr.edu

A project of NASA and the University of California, Riverside, the Alternative Earth's Astrobiology Center studies the possibility of life on exoplanets by studying the bio-signatures of life on our own planet throughout its history.

Astrobiology Initiative

University of California, Santa Cruz
Center for Adaptive Optics
1156 High Street
Santa Cruz, CA 95064
(831) 502-7042
email: dtanguay@ucsc.edu
website: www.astrobiology.science.uscs.edu

The scientists at the UC Santa Cruz's Astrobiology Center seek to answer the question "are we alone in the universe" by seeking signatures of life on other planets.

Australian Centre for Astrobiology

University of New South Wales
Biological Sciences Building (D26)
Gate 9, UNSW
Sydney, NSW 2052
Australia
email: carol.oliver@unsw.edu.au
website: www.aca.unsw.edu.au

The Australian Center for Astrobiology, the only such research center in Australia, is an associate member of the NASA Astrobiology Institute. The ACA does interdisciplinary research into the origin of life on Earth and throughout the solar system.

European Space Agency (ESA)

24 rue du Général Bertrand
CS 30798
75345 Paris CEDEX 7
France
email: education@esa.int
website: www.esa.int

The European Space Agency (ESA) is an international organization of twenty-two member states. The agency is dedicated to the exploration of space for peaceful purposes. It is headquartered in Paris, France.

Goddard Space Flight Center/Astrobiology at GSFC

NASA's Goddard Space Flight Center
Public Inquiries
Mail Code 130
Greenbelt, MD 20771
(301) 286-2000
email via contact page on website
website: https://astrobiology.gsfc.nasa.gov

The Goddard Space Flight Center, established in 1959, is one of NASA's largest space flight centers. GSFC is involved in every aspect of the science of astrobiology.

JAXA (Japan Aerospace Exploration Agency)

Ochanomizu sola city
4-6 Kandasurugadai
Chiyoda-ku, Tokyo 101-8008
Japan
website: www.global.jaxa.jp

Established in 2015, JAXA supports the development and utilization of the Japanese government's aerospace programs.

Kennedy Space Center

Space Commerce Way
Merritt Island, FL 32953
(855) 433-4210
website: www.kennedyspacecenter.com

The Kennedy Space Center is one of NASA's ten field centers. Since 1968, it has been the primary launch center for human spaceflight.

NASA

300 E Street SW, Suite 5R30
Washington, DC 20546
(202) 358-0001
email via contact page on website
website: www.nasa.gov

NASA (National Aeronautics and Space Administration) is the civil space program of the United States. NASA, established in 1957, is the government agency responsible for science and technology related to space.

SETI Institute

189 Bernardo Avenue
Suite 200
Mountain View, CA 94043
(650) 961-6633
email: Info@seti.org
website: www.seti.org

SETI (Search for Extraterrestrial Intelligence) is a nonprofit research organization dedicated to the search for and understanding of life beyond Earth. SETI was established in 1984.

University of Colorado Center for Astrobiology

Laboratory for Atmospheric and Space Physics
12324 Innovation Drive
Boulder, CO 80303
(303) 492-6412
email: epomail@lasp.colorado.edu
website: www.lasp.colorado.edu/home/science/centers/center
-for-astrobiology/

The work of the University of Colorado's Center for Astrobiology includes research, teaching, and community outreach in the area of astrobiology.

Bibliography of Books

Jim Al-Khalili. *Aliens: The World's Leading Scientists on the Search for Extraterrestrial Life*. New York, NY: Picador, 2017.

David C. Catling. *Astrobiology: A Very Short Introduction*. Oxford, UK: Oxford University Press, 2013.

Christian Davenport. *The Space Barons: Elon Musk, Jeff Bezos, and the Quest to Colonize the Cosmos*. New York, NY: Public Affairs, 2018.

Leonard David. *Moon Rush: The New Space Race*. Washington, DC: National Geographic, 2019.

Kevin Hand. *Alien Oceans: The Search for Life in the Depths of Space*. Princeton, NJ: Princeton University Press, 2020.

Elizabeth Howell and Nicholas Booth. *The Search for Life on Mars: The Greatest Scientific Detective Story of All Time*. New York, NY: Arcade, 2020.

Arik Kerschenbaum. *The Zoologist's Guide to the Galaxy: What Animals on Earth Reveal About Aliens—and Ourselves*. New York, NY: Penguin, 2021.

Avi Loeb. *Extraterrestrial: The First Sign of Intelligent Life Beyond Earth*. Boston, MA: Houghton Mifflin Harcourt, 2021.

Rod Pyle. *Space 2.0: How Private Spaceflight, a Resurgent NASA, and International Partners Are Creating a New Space Age*. Dallas, TX: BenBella, 2019.

Carl Sagan. *Contact*. New York, NY: Simon and Schuster, 1985.

Alan Steinfeld. *Making Contact: Preparing for the New Realities of Extraterrestrial Existence*. New York, NY: St. Martin's, 2021.

James Trefil and Michael Summers. *Imagined Life: A Speculative Scientific Journey Among the Exoplanets in Search of*

Intelligent Aliens, Ice Creatures, and Supergravity Animals. Washington, DC: Smithsonian Books, 2019.

Michael Wall. *Out There: A Scientific Guide to Alien Life, Antimatter, and Human Space Travel (For the Cosmically Curious).* New York, NY: Grand Central, 2018.

Ryan S. Walters. *Apollo 1: The Tragedy That Put Us on the Moon.* Washington, DC: Regnery History, 2021.

Jon Willis. *All These Worlds Are Yours: The Scientific Search for Alien Life.* New Haven, CT: Yale University Press, 2016.

Robert Zubrin. *The Case for Space: How the Revolution in Spaceflight Opens Up a Future of Limitless Possibility.* Guildford, CT: Prometheus, 2019.

Index

U

V

W